Manipulating the Mouse Embryo

A LABORATORY MANUAL

Frontispiece: Preimplantation stages in mouse embryogenesis. *(1) Fertilized egg. Note the sperm within the zona pellucida. (2) Two-cell stage. A polar body is still visible. (3) Three-cell stage. Cleavage of the blastomeres is asynchronous. (4) Four-cell stage. (5) Five-cell stage. (6) Eight-cell uncompacted morula. (7) Compacted morula. (8) Early blastocyst. A small blastocoel cavity is present. (9) Expanded blastocyst. (10) Hatched blastocyst. Prior to hatching the blastocoel cavity expands and contracts, possibly because the permeability seal created by the zonular apical tight junctions between the trophectoderm cells is temporarily broken when these cells divide. (Photograph provided by Dr. P.R. Braude, Department of Obstetrics and Gynecology, Cambridge University.)*

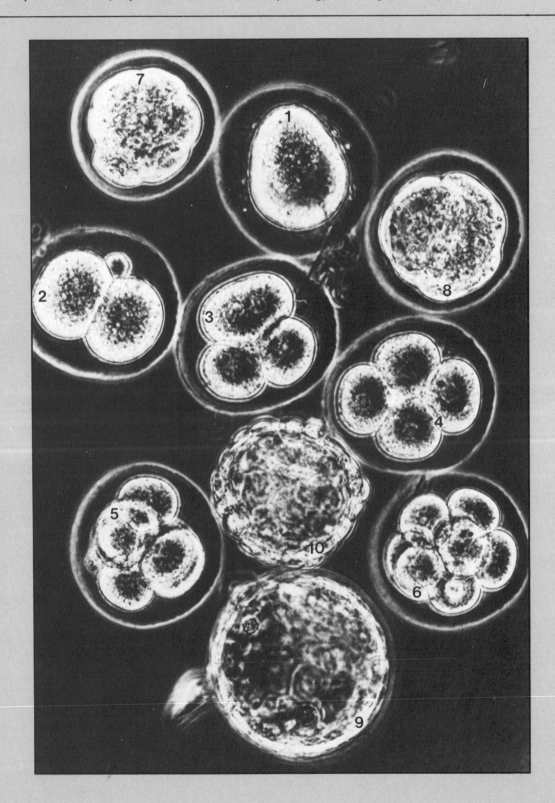

Manipulating the Mouse Embryo

A LABORATORY MANUAL

Brigid Hogan
National Institute for Medical Research

Frank Costantini
Columbia University

Elizabeth Lacy
Memorial Sloan-Kettering Cancer Center

Cold Spring Harbor Laboratory
1986

Manipulating
the Mouse Embryo
A Laboratory Manual

Front cover: A fertilized one-cell mouse embryo, showing the two pronuclei (maternal and paternal) as flattened discs. The embryo is being restrained by a holding pipet, using a slight vacuum on the zone pellucida. A glass capillary needle contains solution of DNA.

Back cover: DNA microinjection of a mouse embryo to produce a transgenic mouse. A fine glass needle is being used to inject a solution of DNA into one of the two pronuclei, which is swelling under the pressure.

Library of Congress Cataloging-in-Publication Data
Manipulating the mouse embryo.
  Bibliography: p.
  Includes index.
    1. Embryology–Mammals. 2. Mice. 3. Molecular biology.
I. Hogan, Brigid. II. Costantini, Frank, 1952-
III. Lacy, Elizabeth.
QL959.M265 1986   599.32'33   84-17628
ISBN 0-87969-175-1

Certain experimental procedures in this manual may be the subject of national or local legislation or agency restrictions. Users of this manual are responsible for obtaining the relevant permissions, certificates, or licenses in these cases. Neither the authors of this manual nor Cold Spring Harbor Laboratory assume any responsibility for failure of a user to do so.

All Cold Spring Harbor Laboratory publications may be ordered directly from Cold Spring Harbor Laboratory, Box 100, Cold Spring Harbor, New York 11724. (Phone: 1-800-843-4388) In New York (516) 367-8423

Preface

These are exciting times for mammalian embryology. The revolution of re-combinant DNA technology has made possible experiments only dreamt of by the pioneers of the field—the ability to isolate and sequence genes, engineer them in specific ways, monitor their expression by in situ hybridization, and rapidly map them by means of restriction fragment length polymorphisms. These feats have been matched by equally impressive advances in techniques for introducing for-eign DNA into the germ line of mice and larger animals. Microinjection of DNA into the fertilized egg, retroviral infection, and transformation of pluripotential embryonic stem cell lines have all opened up ingenious new ways of studying the control of gene expression during development, and of following the conse-quences of altering or blocking specific gene products. Dazzling though these technologies are, they are worth very little unless applied to projects designed to ask fundamental questions about embryonic development. It is here that the mo-lecular biologist can look for inspiration to an impressive tradition in mammalian embryology that has identified and described the major events which need to be understood in molecular terms. The most obvious are cell diversification and dif-ferentiation. But we also know very little about the various morphogenetic pro-cesses which, for example, bring about changes in tissue organization, direct the migration of the germ cells and neural crest, and establish specific nerve connec-tions.

Classical mammalian embryology has also identified many areas highly rele-vant to human reproduction that are ripe for molecular studies, including sex determination, germ cell maturation, implantation, maternal-fetal interaction, and placental function. At another practical level there are now possibilities for in-creasing the yield and potential uses of agricultural animals, and for understand-ing and rectifying inherited defects and childhood cancers.

It is to help catalyze the interaction between molecular biology and mammal-ian embryology that this manual has been written. It grew out of the teaching material for two practical courses in the Molecular Embryology of the Mouse held at Cold Spring Harbor Laboratory in 1983 and 1984. These courses, in turn, could never have been organized without the foresight and enthusiasm of James Wat-son, and we should like to thank him for so much support and inspiration over the last 3 years. It is indeed fitting that Cold Spring Harbor Laboratory was the first to hold courses of this kind. Not only does the Laboratory have a great tradi-tion in molecular biology, but also, as readers of our introduction will learn, it was the birthplace of some of the first inbred mouse strains and the site of pioneering work in mouse genetics.

Many people generously and enthusiastically contributed their hard-won expertise to the courses and the manual. In particular, we should like to express our gratitude to Anne McLaren, who not only proved to be an almost inexhaustible source of information on all aspects of reproductive biology, but also gave gentle sympathy and encouragement to students searching in vain for embryos and oviducts! We also owe special thanks to Christopher Graham, who initially taught us many of the techniques in this manual, and who provided valuable comments on the manual at various stages of its preparation. Lee Silver and Douglas Hanahan generously helped us to organize the animal breeding, laboratory space, and other facilities necessary to run the course, and also provided many helpful comments on the manuscript. We wish to thank the following people for demonstrating techniques in the course and/or contributing information that has been incorporated into the text: Eileen Adamson, Lynne and Bob Angerer, Helene Axelrad, Alan Bernstein, Ralph Brinster, Bruce Cattenach, Verne Chapman, Ted Evans, Susan Howlett, Rudolf Jaenisch, M.H. Kaufman, Robb Krumlauf, Cecelia Lo, James McGrath, Hester Pratt, Liz Robertson, Michael Rosenberg, Janet Rossant, Davor Solter, and Heidi Stuhlman. We also thank Denise Barlow, Kiran Chada, Kathie Raphael, and Liz Robertson, who served as assistant instructors in the course.

Finally, we could not have produced this manual without the skill and expertise of the Cold Spring Harbor Laboratory Publications Department, headed by Nancy Ford. Our editor Judy Cuddihy, in particular, organized much of the production and guided us through times when it seemed that the Manual would never be finished.

B.L.M.H.
F.C.
E.L.

Contents

Manipulating the Mouse Embryo

A LABORATORY MANUAL

Developmental Genetics and Embryology of the Mouse: Past, Present, and Future

Progress in the genetic analysis of mammalian development can never be as rapid as with *Drosophila* and *Caenorhabditis*. Not only is the mammalian genome size larger and the generation time longer, but the embryos also develop much more slowly. In addition, they are adapted to grow within the protective and nutritive environment of the mother, so that it is more difficult experimentally to manipulate mouse embryos compared with those of *Xenopus* or chick. In spite of these drawbacks, there is a unique challenge to understanding how genes control the growth and differentiation of the mammalian embryo. To some extent, this challenge is an intellectual one, and derives from our curiosity to know how human form is generated. But at a practical level we also need to know how mutations and chemicals produce human malformations, congenital defects, and childhood cancers, and whether the productivity of agricultural animals can be improved. This knowledge, and the ability we now have to change the genetic program, must inevitably make a great impact on society and have far-reaching effects on the way in which we think about ourselves.

The roots of our knowledge about how genes control mammalian development can be traced back to experiments carried out in the early 1900s on the inheritance of coat colors in a variety of domestic animals. Since then the mouse has become firmly established as the primary experimental mammal, and more information has accumulated on its genetics than on any other vertebrate, including man. Over 700 genes, many of them affecting development, have been mapped onto its haploid set of 20 chromosomes (see genetic maps in Appendix and sources listed at the end of this chapter), and regions such as the *H-2*, *t*, and *dilute–small ear* complexes have been analyzed in great detail (for review, see Klein 1975; Frischauf 1985; Rinchik et al. 1985; Silver 1986). Techniques of molecular biology, including in situ hybridization, are being applied to the expression of specific genes at different stages of development, and to the isolation of new lineage markers.

Recently, however, a new impetus and sense of excitement have been generated by three important advances. One is the ability to introduce new genetic information into all somatic tissues, as well as the germ line, either by microinjecting DNA into the pronucleus of the fertilized egg or by infecting embryos with retroviral vectors. The second is the technique of transplanting nuclei of fertilized eggs and oocytes. And the third is the ability to establish from normal mouse embryos in culture pluripotential stem cell lines that can be reintegrated back in

1

the embryo by injection into host blastocysts. Transforming these cells with DNA before injection may be another route into the germ line.

The aim of this book is to provide a simple technical manual for scientists wanting to learn these new techniques. This manual also describes other well-established procedures for manipulating the mouse embryo, and includes a summary of early mouse embryogenesis. We hope that in making this information available to a wide audience the manual will help to continue the spirit of cooperation established by the first mouse geneticists.

Mendelian Inheritance and Linkage: The Beginnings of Mouse Genetics

Historians of science on both sides of the Atlantic acknowledge the American scientist William E. Castle as one of the founding fathers of mammalian genetics. As first director of the new Bussey Institute of Experimental Biology in Harvard, from 1909 to 1937, he encouraged work on the inheritance of variable characters in a wide range of organisms, including birds, cats, dogs, guinea pigs, rabbits, and rats, as well as mice (Russell 1954; Keeler 1978; Morse 1978, 1981). He was even responsible for introducing Thomas Hunt Morgan to *Drosophila* (Shine and Wrobel 1976). Through the many scientists that came to visit or study at the Bussey Institute, Castle had a profound influence on the course of mammalian genetics.

Of the diversity of mammals studied by these early geneticists, it soon became clear that the mouse had the advantage of small size, resistance to infection, large litter size, and a relatively rapid generation time (see Table 1). The use of mice was also favored by the interesting pool of mutations affecting coat color and behavior readily available from breeders and collectors of pet mice, or mouse "fanciers." One of these mutants, albino (see Fig. 30), was used by Bateson in England, Cuenot in France, and Castle in the United States for the first breeding experiments demonstrating Mendelian inheritance in the mouse (see references in Castle and Allen 1903). A few years later, albino and another old mutation of the mouse fanciers, pink-eyed dilute (Fig. 30), were used by J.B.S. Haldane for the first demonstration of linkage in mice (Haldane et al. 1915). Sadly, this work was interrupted in 1914 when Haldane volunteered for service in the First World War, leaving his sister to continue their experiments for a while in the Department of Comparative Anatomy in Oxford (Clark 1984; Naomi Mitcheson, pers. comm.). It was not until after the war that Haldane was able to turn his attention to the wider aspects of mammalian genetics and, along with others, begin developing mathematical models of inheritance and natural selection.

Origins of the Laboratory Mouse

If William E. Castle and J.B.S. Haldane are founding fathers of mouse genetics, then the mother is undoubtedly Abbie E.C. Lathrop. A self-made woman, Abbie Lathrop established around 1900 a small mouse "farm" in Granby, Massachusetts, to breed mice as pets. However, her mice were soon in demand as a source of experimental animals for the Bussey Institute and other American laboratories, and she gradually expanded her work to include quite sophisticated and well-documented breeding programs. For example, in collaboration with Leo Loeb she

Table 1 Some Vital Statistics of the European House Mouse, *Mus musculus*, in the Laboratory

Genome	
Number of chromosomes	40
Diploid DNA content	~6 pg (3×10^9 bp)
Recombination units	1600 centimorgans (2000 kb/cM)
Approximate number of genes[a]	$0.5–1.0 \times 10^5$

Percent of genome as five
      families of highly repeated
      DNA sequences (B1, B2, R,
      MIF-1, and EC1)[b]
	8–10%

Reproductive biology[c]	
Gestation time	19–20 days
Age at weaning	3 weeks
Age at sexual maturity	~6 weeks
Approximate weight	birth 1 g weaning 8–12 g adult 30–40 g (male > female)
Lifespan in laboratory	1.5–2.5 years
Average litter size[d]	~6–8
Total number of litters per breeding female	4–8

[a]McKusick and Ruddle (1977).

[b]Bennett et al. (1984).

[c]Parameters such as gestation time, weight, lifespan, etc., vary between the different inbred strains. Details can be found in a number of books listed in Sources of Information on Genetic Variants and Inbred Strains of Mice and Their Genetic Monitoring (at end of this section), e.g., Altman and Katz (1979), Festing (1979), and Heiniger and Dorey (1980).

[d]Litter size depends on the number of eggs liberated at ovulation and the rate of prenatal mortality, both of which vary with age of mother, parity, and environmental conditions (e.g., diet, stress, presence of strange male) and with strain (reflecting genetic factors such as efficiency of placentation). Prenatal mortality in inbred strains can be around 10–20% (for references, see Boshier 1968).

carried out experiments to study the effect of genetic background, inbreeding, and pregnancy on the incidence of spontaneous tumors in her mice (Shimkin 1975; Morse 1978). As source material for the farm, Abbie Lathrop used wild mice trapped in Vermont and Michigan, fancy mice obtained from various European and North American sources, and imported Japanese "waltzing" mice. Waltzing mice had been bred as pets in China and Japan for many generations and were probably homozygous for a recessive mutation that causes a defect in the inner ear and thus nervous, circling, behavior. The Granby mouse farm was, to a large extent, the "melting pot" of the laboratory mouse, and, as shown in Table 2, many of the old inbred strains can be traced back to the relatively small pool of founding mice that Lathrop maintained there. The genealogies of other common laboratory strains can be found in several sources (e.g., Morse 1978, 1981).

The full significance of these lineages is only now being revealed by the application of restriction fragment length polymorphism (RFLP) studies to mouse DNA.

Table 2 Genealogy of the More Commonly Used Inbred Mouse Strains

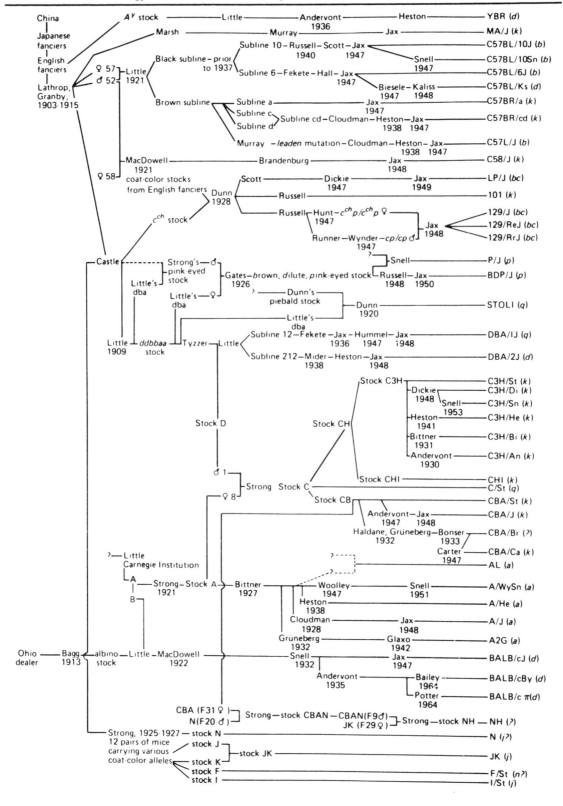

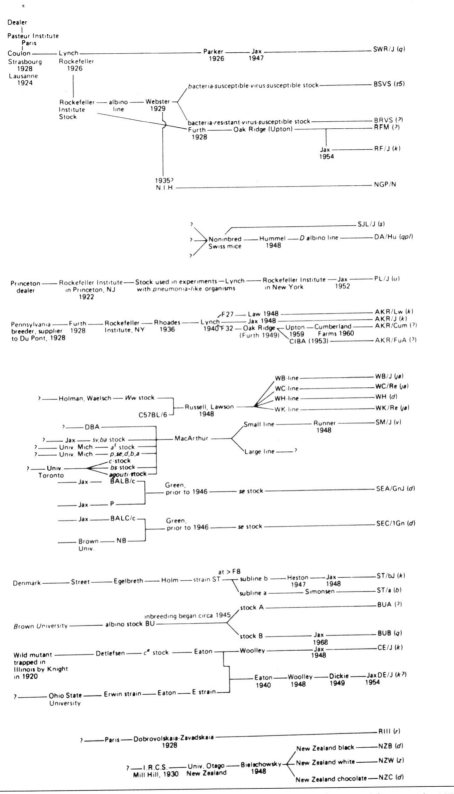

Table is based, in part, on data provided by Michael Potter and Rose Lieberman in 1967; it was extended by Jan Klein in 1975; and it was revised by Potter in 1978. *H-2* haplotypes are shown in parentheses.

Table is reproduced, with permission, from Altman and Katz (1979).

Analysis of the RFLP of mitochondrial DNA (which is maternally inherited through the egg cytoplasm) has shown few differences among old established strains, compared with the wide variations seen among wild mice and newer strains derived from them. In fact, on the basis of mitochondrial DNA RFLPs, it has been argued that at least five of the primary strains (DBA, BALB/c, SWR, PL, and C57-C58) were derived originally from a single female of the *Mus musculus domesticus* group (Ferris et al. 1982). This group is found in western and southern Europe and is the source from which all wild mice in the northern parts of the United States were derived by migration with humans across the north Atlantic shipping lanes. A second taxonomic group, *Mus musculus musculus*, is found in central and eastern Europe, Russia, and China, and only interbreeds with *domesticus* over a narrow band from north to south through central Europe (Bonhomme et al. 1984) (Fig. 1). In addition to having distinct mitochondrial DNA RFLPs, the two groups also show different patterns using DNA probes specific for the Y chromosome, which is inherited only through the male (Bishop et al. 1985; Bulfield 1985). Unexpectedly, in view of the mitochondrial RFLP data, many old inbred mouse strains, including A/J, BALB/c, C57BL/6, CBA/HeJ, C3H, DBA/2, 129/Sv, and 163/H, have Y chromosome RFLPs of the *musculus* type. The most likely explanation of this is that the Y chromosome came from Japanese pet mice, for example, those bred on the Granby mouse farm (Bishop et al. 1985).

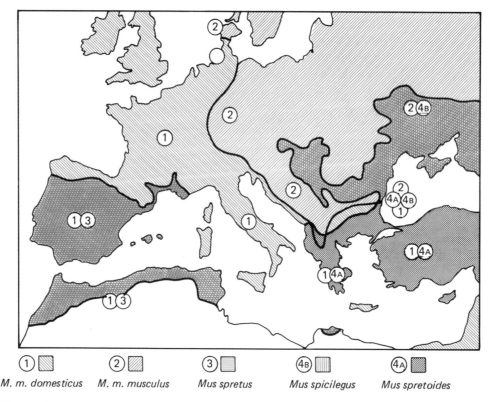

①	②	③	④B	④A
M. m. domesticus	*M. m. musculus*	*Mus spretus*	*Mus spicilegus*	*Mus spretoides*

Figure 1 Geographical distribution of the five biochemical groups of the house mouse species complex in Europe. (Redrawn from Bonhomme et al. 1984.)

Origin of Inbred Strains and Other Resources of Mouse Genetics

An inbred strain is defined as one that has been maintained for more than 20 generations of brother-to-sister mating and is essentially homozygous at all genetic loci, except for mutations arising spontaneously (Altman and Katz 1979; Morse 1981). The derivation of inbred strains represents one of the most important phases in the history of mouse genetics and it revolutionized studies in cancer research, tissue transplantation, and immunology. One of the pioneers of the innovation was Clarence C. Little. He was originally a student of Castle's at the Bussey Institute, where he studied the inheritance of mouse coat color, and he later went on to found the Roscoe B. Jackson Memorial Laboratory (otherwise known as the Jackson Laboratory) in Bar Harbor, Maine (Russell 1978; Morse 1981). Other pioneers were Lionelle Strong, Leo Loeb, and Jacob Furth (Morse 1978; Strong 1978). Among the first inbred strains were DBA, which was named after the coat color mutations it carried—dilute (*d*), brown (*b*), and nonagouti (*a*), A, and C57 and C58, which were derived from females 57 and 58 from the Granby mouse farm. While carrying out these early inbreeding experiments, both Little and Strong worked between 1918 and 1922 at the Carnegie Institute of Washington at Cold Spring Harbor, thus establishing the laboratory (then known as the Station for Experimental Evolution) as one of the birthplaces of mouse genetics (Keeler 1978; Strong 1978).

In deriving the strains, great tenacity was required to maintain the strict brother-to-sister matings through times when the breeding stocks reached a very low ebb due to disease or accidents, and accounts of these difficult times make fascinating reading (Morse 1978). It also required intellectual courage to challenge the widely held belief that inbreeding to virtual homozygosity would be impossible due to recessive lethal mutations in the founding pairs. Today, almost 200 inbred mouse strains are available. Each has a standardized nomenclature, to indicate strain and substrain, and details of their history and characteristics are available from the several sources listed at the end of this chapter. Standard methods for maintaining breeding colonies and testing mice for genetic purity are described in a recent publication, also listed at the end of this chapter (Nomura et al. 1985). Newcomers to the field should be aware that examples of cross-contamination of stocks from commercial sources are by no means rare, even in the 1980s.

One of the driving forces behind the initial establishment of inbred strains was the need to rationalize studies on the genetics of cancer susceptibility. Inbred strains were also essential for solving the problem of why spontaneous tumors could be transplanted into some mice and not others. Although many groups studied this problem, a major contribution was made by Peter Gorer, working in Haldane's department in University College, London. Using A, C57BL, and DBA strains of mice and a transplantable A strain tumor, he showed for the first time that mice resistant to tumor growth produced antibodies against antigens present not only on the tumor cells but also on blood cells of A strain mice. One particularly strong antigen was called Antigen II. In 1948, Gorer and the American geneticist George Snell together showed that the gene specifying Antigen II was closely linked to the fused (*Fu*) locus (now known to be on chromosome 17), and they called the gene *Histocompatability-2*, or *H-2* (Gorer et al. 1948). In a series of outstanding experiments, for which he was awarded the Nobel Prize in 1980, Snell went on to identify and map many of the minor histocompatability loci as well. All of this work was carried out at the Jackson Laboratory and owes much to the unique environment built up there by C.C. Little and his colleagues. It was the first laboratory in which many inbred strains were maintained under conditions of strict breeding and health monitoring, and from the time of its foundation a spirit of cooperation prevailed (Morse 1978; Russell 1978; Snell 1978).

To identify the histocompatibility genes, Snell developed the concept of congenic inbred strains, in which short segments of the chromosome around a marker gene were transferred from one strain into an inbred genetic background by repeated backcrossing and selection. Like the inbred strains, congenic strains have a strict nomenclature (Snell 1978; Altman and Katz 1979; Morse 1981). For example, B10.129-H-12^b is a strain in which the allele (*H-12^b*) derived from the strain 129/J has been transferred onto the C57BL/10 inbred background. Many of the congenic strains originally developed by Snell and subsequently by others are widely available from commercial sources, including the Jackson Laboratory.

Another important innovation in mouse genetics was the development of recombinant inbred strains by Donald W. Bailey and Benjamin A. Taylor (Morse 1981). These strains were derived by crossing two highly inbred (progenitor) strains and then inbreeding random pairs of the F_2 generation to produce a series of recombinant inbred or RI strains (Table 3). Their usefulness is in localizing within chromosomes any new locus that shows a polymorphism between the two progenitor strains. This is done by comparing the strain distribution pattern (SDP) of the new polymorphism with the many SDPs already established for enzyme, protein, or DNA RFLPs associated with known loci. The advantage of the system is that the data are cumulative; the patterns already published (Green 1981), or stored on computer at the Jackson Laboratory, provide a unique data base for mapping the mouse genome. The exponential increase in the number of cDNA probes for known proteins has produced a corresponding escalation in chromosomal assignments. One disadvantage of RI strains is that they are expensive to maintain, but DNA is now available from the Jackson Laboratory. Another disadvantage is the difficulty

Table 3   Schematized Construction of Eight RI Strains

Progenitor inbred strains				AABBCC × aabbcc			
↓							
F_1				AaBbCc			
↓							
F_2				AaBbCc × AaBbCc			
↓							
Inbreeding for more than 20 generations							

RI Strain	1	2	3	4	5	6	7	8
	AABBCC	AABBcc	AAbbCC	AAbbcc	aaBBCC	aaBBcc	aabbCC	aabbcc
A	A	A	A	A	a	a	a	a
B	B	B	b	b	B	B	b	b
C	C	c	C	c	C	c	C	c

Construction starts from two progenitor strains that have alternate alleles at three unlinked loci. The three alleles segregate and assort independently during the inbreeding process and eventually become genetically fixed. Each allele then has a unique strain distribution pattern (SDP).

often encountered in finding polymorphisms among the progenitor strains. This is due in part to the rather restricted origin of laboratory mice, as discussed in the previous section. Recently, an alternative mapping technique has been described based on backcrossing an F_1 hybrid between an inbred strain and *M. spretus*, a wild mouse species found in Spain. Because *M. m. domesticus* and *M. spretus* are different species, the chances of finding an RFLP for any given DNA probe are much higher (Robert et al. 1985).

Wild mice have contributed to laboratory studies in other ways. For example, as shown originally by the German geneticist Alfred Gropp, they can be used to introduce cytogenetic variations into the karyotype of *M. m. domesticus*, which otherwise consists of 40 acrocentric chromosomes that are very difficult to distinguish (see Karyotyping of Mouse Cells, Section F). Gropp discovered in high Swiss valleys inbred groups of mice that have seven pairs of bi-armed (or Robertsonian fusion) chromosomes produced by the centric fusion of pairs of normal chromosomes (Gropp and Winking 1981). Individual Robertsonian chromosomes have been crossed into inbred laboratory strains where they can be used to generate embryos that are monosomic or trisomic for particular chromosomes (Epstein 1985). They also provide markers for genetic or cytogenetic experiments including the mapping of genes by in situ hybridization (e.g., Munke et al. 1985). Since their discovery in mice of the Valle di Poschiavo, centric fusions have been found in other localities, and in laboratory strains. Like inbred strains, they have a strict nomenclature; for example, Rb (11.16)2H is a Robertsonian fusion involving chromosomes 11 and 16 and was the second of a series identified at the MRC Radiobiology Laboratory at Harwell (H) (Green 1981).

Another example of the use of the genetics of wild mice has been the recent introduction of protein polymorphisms at the PGK and HPRT loci into congenic strains for studies on X chromosome inactivation (Nielsen and Chapman 1977; Chapman et al. 1983). For more information on the genetics and natural history of *M. m. domesticus* and its relatives, the reader is referred to an excellent symposium volume *Biology of the House Mouse* (Berry 1981).

Origins of Developmental Genetics of the Mouse

Because of their availability from the mouse fancy, many of the first mutants used in breeding experiments sported visible differences in coat color, hair morphology, and pigmentation patterns (for example, Fig. 30). Even today, these mutations present a virtually untapped resource for studies on many cell biological problems, including the structure of intracellular organelles (e.g., the beige mutation, see Fig. 30), cell migration (e.g., extreme white spotting, Fig. 30), tissue interactions (e.g., agouti and belted, Fig. 30), and unstable genes and position-effect variegation (e.g., chinchilla mottled, Fig. 30).

Over the years, other mutants affecting more complex neurological, physiological, and morphogenetic processes have been identified. Some were uncovered during the early days of inbreeding as recessive mutations in wild or fancy mice. Others have arisen as spontaneous mutations in laboratory stocks of already inbred mice. Another important source has been the offspring of mice exposed to X-rays or chemical mutagens. Much of this work has been carried out in two laboratories established shortly after the Second World War in response to the need for research into the biological effects of radiation: the Oak Ridge National Laboratory, Tennessee, and the MRC Radiobiology Unit in Harwell, near Oxford. As well as generating a whole range of important radiation-induced mutants and chromosomal rearrangements, these laboratories have done outstanding work on basic mouse genetics. For example, in Oak Ridge, Liane Russell has mapped overlapping deletions covering the *dilute–short ear* region on chromosome 9 which includes several genes involved in prenatal development (Russell 1971; Rinchik et al. 1985). On the other side of the Atlantic, Mary Lyon, working first at the Department of Genetics in Edinburgh and then at Harwell, was the first to describe the phenomenon of random X chromosomes in somatic tissues of female mice (Lyon 1961). At Harwell she has also generated many new ideas about the genetic organization of the *t* complex, which will be discussed later (Lyon et al. 1979). By whatever route mouse mutants and chromosomal variants are derived, they are very expensive and time consuming to isolate and maintain, and those that have been conserved and are catalogued in the sources listed at the end of this chapter are testament to an enormous amount of hard work, dedication, and foresight by many mouse geneticists.

Looking back, it is also easy to underestimate the painstaking work that went into describing the pathology and etiology of many of the early morphological mutants. It soon became apparent that to understand how a whole range of defects in the adult could be caused by mutation in a single gene, it was necessary to trace the mutant phenotype back into the early embryo. One geneticist who made a speciality of this approach was Hans Grüneberg, a refugee from Germany, who in 1938 was invited by Haldane to work at University College, London. Originally a physician, Grüneberg was motivated by a belief that mouse mutants could be used as models for understanding human congenital defects. For more than 40 years he described a whole variety of mutants, in particular those with skeletal abnormalities (see, for example, pudgy, Fig. 30). He traced many of them back to early postimplantation stages when they first showed signs of defects in the process of somite formation and differentiation. His books, *Animal Genetics and Medicine* (1947), *The Genetics of the Mouse* (1952), and *The Pathology of Development* (1963), are classics of their kind and were as influential as Ernst Hadorn's *Developmental Genetics and Lethal Factors*, which was written in 1955 and translated into English in 1961.

Foremost among the pioneers of mouse developmental genetics in the United States are L.C. Dunn, a contemporary of Thomas Hunt Morgan at Columbia University, and his colleagues Salome Gluecksohn-Waelsch, originally a student of Spemann's, and Dorothea Bennett. Dunn and his disciples can be credited with describing many homozygous lethal mouse mutants, but their most significant contribution has been to promote the genetic analysis of the *t* complex on chromosome 17. The first mutant forms of this complex were discovered by a Russian cancer research scientist, Nelly Dobrovolskaia-Zavadskaia, working in the Pasteur Institute in Paris on the effects of radiation. She found that one of the offspring of an x-irradiated mouse had a short tail, and defined it as having a dominant mutation, *T*. Later, Dobrovolskaia-Zavadskaia crossed one of her *T* mice with a wild mouse she had caught and was surprised to find that the offspring had no tails at all. In fact, we now know that the wild mouse carried a mutant form of the *t* complex which is called a *t* haplotype and which interacts with *T* to

produce taillessness. In 1932, realizing the complexity of the system she had uncovered and the limitations of her own resources, Dobrovolskaia-Zavadskaia
passed her mice on to Dunn.

Soon a number of different *t* haplotypes were discovered in wild mouse populations, and it was shown that embryos homozygous for different *t* haplotypes
died at different stages of development. The recessive lethal mutations were maintained at high levels in the wild because males heterozygous for wild-type and *t*-
haplotype forms of chromosome 17 transmitted the latter to more than 90% of
their offspring, a phenomenon known as transmission ratio distortion. A widely
held view was that the *t* complex would turn out to encompass a multigene family
encoding stage and tissue-specific surface proteins involved in cell–cell recognition
or interaction during embryogenesis.

Unfortunately, 50 years on, there is little evidence to support this view. It has
emerged that the *t* complex in fact covers over 16 cM of DNA, or about one-third
of chromosome 17 (equivalent to 1% of the entire genome), and contains two large
inversions, one of which includes the *H-2* complex (Herrman et al. 1986). It seems
very likely that these inversions trap lethal mutations by inhibiting recombination
between wild-type and mutant chromosomes, and the complex is further held
intact by transmission distortion (for reviews, see Lyon et al. 1979; Silver 1981,
1986; Frischauf 1985). There is no evidence to suggest that the 16 *t*-haplotype-
associated lethal mutations so far identified are other than genes coding for cell
products that can become rate-limiting at certain stages of development. A useful
analogy is the way in which an insertional mutation in the $\alpha 1(\mathrm{I})$ collagen gene can
cause embryonic death of homozygotes at around 11.5 days p.c. (Lohler et al.
1984); although absence of type-I collagen is incompatible with normal development, the gene itself would not normally be classed as a developmental one. Seen
from this point of view, analysis of the genes responsible for *t*-haplotype effects
will undoubtedly throw some light on embryogenesis, spermatogenesis, and the
mechanism of transmission ratio distortion, and some genes may turn out to have
important regulatory functions. But there is no reason to believe that most of the
DNA involved in the *t* complex is uniquely involved in embryonic development.

It is hard to escape the general conclusion that in spite of all the work that has
gone into the phenotypic analysis and breeding of mouse mutants over the years,
very little has so far been learned from them about the genetic control of differentiation and morphogenesis. In a few cases, a direct link has been established
between a mutant phenotype and a defective protein product (e.g., a defect in
basic myelin protein and the shiverer mutant; Roach et al. 1985). But there is still
no easy way of homing in on mutations and deletions affecting interesting developmental processes and identifying the genes involved.

There are several ways in which this dilemma may be resolved in the future.
For one thing, as more and more cloned genes are mapped, some may, by chance,
fall near developmental loci and provide a starting point for genomic "walking."
Also, by chance, DNA or a retrovirus used to generate transgenic mice may insert
into a gene controlling a specific developmental decision or morphogenetic process. The foreign DNA will then provide a unique handle for isolating the controlling gene. An example of such insertional mutagenesis has recently been described
(Woychik et al. 1985), although the nature of the gene mutated is not yet known.
On another tack, mouse homeobox-containing genes may turn out to be selector

genes with regulatory functions (for review, see Hogan et al. 1985) and surveys for other evolutionary conserved sequences may establish yet more unexpected links between *Drosophila* and mammals. The technique of microcloning used successfully for the *t* complex and the X chromosome (Fischer et al. 1985) may be applied to other chromosomal regions encompassing known developmental mutations. Differential screening of cDNA libraries of gonadal tissues may be used to isolate genes involved in sexual differentiation (A. McLaren, pers. comm.). Finally, working in the other direction, studies in experimental embryology will soon identify proteins involved in embryonic induction, cell–cell adhesion and interaction, and pattern formation (see, for example, Edelman 1985; Hatta and Takeichi 1986; Rutishauser and Goridis 1986). Once the genes for these proteins have been cloned, transgenic mice will provide a unique test-bed for theories about their function and evolution.

Origins of Experimental Mouse Embryology

Omne vivum ex ovo. Harvey, 1651

Mammalian genetics had a clear beginning with the rediscovery of Mendel's laws in 1900, and was initially championed by a relatively small number of enthusiasts, centered around leaders like Castle, Little, and Haldane. In contrast, mammalian embryology is a much older science, and it would be beyond the scope of this Introduction to trace the complex lineage of modern ideas back through many communities and continents to the classical experimental embryologists like Boveri, Roux, Spemann, Hadorn, Nieuwkoop, and Waddington. Also, from the beginning, mammalian embryology was closely associated with studies into human and veterinary reproductive physiology, and it is through these links that social pressures have had their influence on academic research. For example, the accelerated pace of research into mammalian reproduction and embryology in the late 1950s and 1960s was due, in part, to the realization of the need for new methods of human population control and increased food production. Many laboratories in the United States and Europe were funded by the Population Council, Inc., and by the Ford Foundation. Other support came from bodies like the Agricultural Research Council in Great Britain who were anxious to see improvements in the fertility and yields of farm animals. Readers interested in the history of mammalian embryology will find references in Needham (1959), Oppenheimer (1967), Mayr (1982), and Austin (1961).

Much of the early experimental work in mammalian embryology was done using rabbit embryos. This included accurate description of preimplantation stages (Van Beneden 1875), oviduct transfer (Heape 1890), filming of morulae dividing in culture (Lewis and Gregory 1929), and other in vitro observations (for reviews, see Pincus 1936; Austin 1961). Rabbits were used initially because the eggs are relatively large and easy to handle, being surrounded by a thick mucin coat, and the female ovulates only after mating, so that the age of the embryos could be timed quite precisely. However, these advantages were soon outweighed as more became known about the reproductive physiology and genetics of mice.

The first report of attempts to culture mouse embryos in vitro to the blastocyst stage was by John Hammond, Junior, son of his namesake the great animal husbandry scientist who introduced the technique of artifical insemination for cattle. Working at the Strangeways Laboratory in Cambridge, England, Hammond Junior succeeded in culturing eight-cell morulae to blastocysts, but embryos removed at two-cell-stage soon died (Hammond 1949). It was not until 1956 that a breakthrough was made by an Australian veterinary scientist Wesley Whitten working in the Australian National University in Canberra after training in Oxford. The motivation behind these experiments was to obtain a defined medium in which the possible requirement of steroid hormones for embryo development could be tested. Using a Krebs-Ringer's bicarbonate solution supplemented with bovine serum albumin, Whitten succeeded in culturing one-cell mouse eggs to the blastocyst stage (Whitten 1956).

Whitten later emigrated to the United States and continued his work in the Jackson Laboratory, but he also collaborated closely with John Biggers of the University of Pennsylvania in Philadelphia. It was in Biggers' laboratory that another veterinarian, Ralph Brinster, began his research career by defining the precise nutritional requirements of the preimplantation mouse embryo and, in the process, established the microdrop culture technique (for review of early work, see Brinster 1965; Whitten and Biggers 1968; Biggers et al. 1971).

These culture conditions, although in the end simple enough, opened up a whole new range of experiments. At the same time, the work of Anne McLaren in the United Kingdom on optimizing conditions for oviduct and uterine transfer (McLaren and Michie 1956) made it possible to overcome the final hurdle, and routinely turn cultured eggs into live mice (McLaren and Biggers 1958). Together, these technical improvements meant that it was at last feasible to test the end result of experimental manipulations on large numbers of embryos. For example, Kristof Tarkowski in Warsaw was able to start analyzing the developmental potential of single mouse blastomeres, using the classical embryological approach of killing one blastomere and seeing how the other would develop. He was also able to make the first aggregation chimeras, an idea conceived and accomplished during a visit to the University of Bangor in north Wales (Ian Wilson, pers. comm.). Tarkowski's original method involved breaking the zona pellucida mechanically and pushing the embryos together in a small drop of medium, which was technically extremely difficult. The whole process was made much easier by Beatrice Mintz in Philadelphia, who discovered that the zona could be gently digested by Pronase. So here again, a relatively simple procedure, once established, opened up a rich mine of biological problems that could be tackled experimentally.

Chimeras of two or more genotypes have been used to study such diverse topics as pigment patterns, sex determination, germ cell differentiation, immunology, tumor clonality, size regulation, and cell lineage (for review, see McLaren 1976). In vitro culture also led to the development of routine methods for parthenogenetic activation of mouse oocytes, and Ralph Brinster was able to carry out the first experiments on the injection of purified globin mRNA into mouse eggs (Brinster et al. 1980).

As far as studies on the postimplantation mouse embryo are concerned, there was considerable debate and confusion about the lineage of the different embryonic and extraembryonic tissues. The various conflicting theories have been summarized by Rossant and Papaioannou (1977). To resolve these problems, and to ask when early embryonic cells become committed to their developmental fate, Richard Gardner in Cambridge, England, developed the technique of generating chimeras by injecting isolated cells into host blastocysts (Gardner 1968). To test the developmental potential of different parts of the postimplantation embryo, several laboratories also developed methods for culturing isolated pieces of tissue in vitro and in ectopic sites. In this way Nikola Skreb and his colleagues in Zagreb showed that the early embryonic ectoderm contains cells capable of contributing to all three germ layers of the fetus.

These studies on the pluripotentiality of cells from the normal embryo were complemented by the use of teratocarcinomas as a model system for studying early embryonic development, an approach pioneered by Leroy Stevens in the Jackson Laboratory and by Barry Pierce in the University of Colorado. Stevens first observed that male mice of the inbred 129 strain have a low incidence of testicular teratoma (Stevens and Little 1954), and then went on to identify genes that increased this frequency and eventually developed a strain (129/Sv) in which the incidence is as high as 30%. Genital ridges from these (and other) mice develop into teratomas or teratocarcinomas when grafted into ectotopic sites. Stevens also developed the LT strain in which about 50% of females develop ovarian teratocarcinomas.

The availability of transplantable teratocarcinomas was another technical advance in mammalian embryology that soon led to many new experiments. It was not long before it was shown by Boris Ephrussi in France and Gordon Sato in the United States, for example, that cells from the tumors could be grown in vitro as cultures that consisted of both differentiated derivatives and undifferentiated stem cells known as embryonal carcinoma. The potential of this culture system for studying the biochemistry and molecular biology of early mammalian embryonic cells was also recognized by François Jacob, and through his influence and the work of his research group in the Pasteur Institute many cell biologists and biochemists were attracted to the teratocarcinoma system and the study of mouse developmental genetics (Jacob 1983).

Finally, the availability of teratocarcinomas led to the demonstration by Brinster (1974), Mintz and Illmensee (1975), and Papaioannou et al. (1975) that embryonal carcinoma stem cells could be reintegrated into blastocysts and contribute to many normal adult tissues. However, it was not until the development of euploid blastocyst-derived cell lines, independently by Gail Martin in San Francisco (Martin 1981) and by Martin Evans and Matt Kaufman in Cambridge, England (Evans and Kaufman 1981), that integration of cultured cells into the germ line could be achieved with high efficiency and reproducibility (for review, see Silver et al. 1983). These cells now represent an important route by which new genetic material may be indirectly introduced into the mouse germ line.

The first report of the direct introduction of new genetic material into the mouse embryo actually predates the widespread use of recombinant DNA techniques. In 1974, Rudolf Jaenisch and Beatrice Mintz found that when purified SV40 viral DNA was injected into the blastocoel cavity of mouse blastocysts, viral DNA sequences could be found in somatic tissues of many of the resulting animals, suggesting that they had integrated into the genome of embryonic cells. In addition, Jaenisch (1976) discovered that Moloney murine leukemia virus could be stably introduced into the germ line by viral infection of preimplantation mouse embryos. However, these studies did not immediately lead to attempts to introduce cloned eukaryotic genes into the germ line.

In 1980, it was reported that the microinjection of the cloned herpes simplex virus (HSV) thymidine kinase (*tk*) gene into the nuclei of cultured fibroblasts led to the stable incorporation and expression of the *tk* gene in 5–20% of the recipient cells (Capecchi 1980; Anderson et al. 1981). This finding suggested that the microinjection of DNA into the one-cell mouse embryo might allow the efficient introduction of cloned genes into the developing mouse and led a number of investigators to test this possibility. The first successful introduction of a cloned gene into mouse somatic tissues by pronuclear injection was reported by Gordon et al. (1980). Shortly thereafter, several groups were successful in introducing cloned genes into somatic tissues as well as into the germ line by this technique (Brinster et al. 1981b; Costantini and Lacy 1981; Gordon and Ruddle 1981; Harbers et al. 1981; E. Wagner et al. 1981; T. Wagner et al. 1981). The structure, inheritance, and expression of foreign genes in transgenic mice, and the applications of this technique for the study of mouse development are discussed in Section D.

SOURCES OF INFORMATION ON GENETIC VARIANTS AND INBRED STRAINS OF MICE AND THEIR GENETIC MONITORING

1. Green, M.C. 1981. *Genetic variants and strains of the laboratory mouse.* Gustav Fischer Verlag.

2. McLaren, A. 1976. Genetics of the early mouse embryo. *Ann. Rev. Genet.* 10: 361–388.

3. Foster, H.L., J.D. Small, and J.G. Fox, eds. 1981. *The mouse in biomedical research. Vol. I. History, genetics and wild mice.* Academic Press, New York.

4. Staats, J. 1980. Standardized nomenclature for inbred strains of mice: Seventh listing. *Cancer Res.* 30: 2083–2128.

5. Altman, P.L. and D.D. Katz, eds. 1979. *Inbred and genetically defined strains of laboratory animals. Part I: Mouse and rat.* Federation of American Societies for Experimental Biology. Bethesda, Maryland.

6. Nomura, T., K. Esaki, and T. Tomita. 1985. *ICLAS manual for genetic monitoring of inbred mice.* University of Tokyo Press, Tokyo.

7. Festing, W.F.W. 1979. *Inbred Strains in biomedical research.* Macmillan Press, London.

8. Peters, J., ed. *Mouse News Letter.* This informal, semiannual publication can be obtained from Dr. Peters, MRC, Radiobiology Unit, Chilton, Didcot, Oxon OX11 ORD, England, or from the Jackson Laboratory, Bar Harbor, Maine 04609. In 1986 it is to be published by Oxford University Press. In addition to listings of genetic variants and mouse stocks held by laboratories throughout the world, *Mouse News Letter* contains short reports of new mutants, linkage assignments, and recent experimental results.

9. Heiniger, H.-J. and J.J. Dorey. 1980. *Handbook on genetically standard Jax mice.* The Jackson Laboratory, Bar Harbor, Maine.

10. Sidman, R.L., M.C. Green, and S.H. Appel. 1965. *Catalog of the neurological mutants of the mouse.* Harvard University Press, Cambridge.

11. Le Douarin, N. 1982. *The neural crest.* Cambridge University Press, England.

12. Daniel, J.C., ed. 1971. *Methods in mammalian embryology.* W.H. Freeman, San Francisco.

Section A

SUMMARY OF MOUSE DEVELOPMENT

In this section we provide a brief survey of mouse development for readers who are completely new to mammalian embryology. For detailed accounts, readers should consult some of the excellent textbooks listed in the Suggested Reading section. We also provide a very short section on the genetics of coat color for people who become interested in this topic after handling mice for the first time. Finally, a few normal and mutant mice are illustrated to give readers at least some idea of the genetic variation that exists in the laboratory mouse.

EARLY MOUSE DEVELOPMENT

Embryonic development of the mouse begins with fertilization of the egg by the sperm. One important feature of mouse embryogenesis is that early development is much slower than in organisms such as the sea urchin, *Drosophila*, and *Xenopus*. By 24 hours after fertilization embryos of these species are well on the way to being free-living, feeding larvae and contain more than 60,000 cells, organized into many different tissue layers. In contrast, the mouse embryo is still at the two-cell stage, and will continue to divide slowly without any increase in mass as it moves along the oviduct into the uterus for implantation 4.5 days after fertilization. This slow development means that the uterine tissue has time to prepare for receiving the embryo. The embryo, in its turn, generates the first two cell lineages (the trophoectoderm and the primitive endoderm) which form the basis of the placenta and the extraembryonic yolk sacs required for successful interaction with the mother. Once implantation has been achieved there is a dramatic increase in the growth rate of the embryo, particularly in the small group of pluripotential cells known as the primitive ectoderm, from which the fetus will develop. The primitive ectoderm is in many ways the equivalent of the cellular blastoderm of *Drosophila*, or the blastodisc of the chick. Between the 5th and 10th day after fertilization the basic body plan of the mouse is established within the cells of the primitive ecto-derm and their descendants. To summarize briefly, the mesoderm is formed and divided up into reiterated pairs of somite blocks, generating a segmented pattern along the anterior–posterior body axis. The neural plate is induced and folds up into the neural tube, and the placodes of the nose, ear, and eyes are formed. The neural crest cells start their migration, and the heart and circulatory system and limb buds are established. Therefore, it is during this period that many of the genes controlling the differentiation and morphogenesis of the adult organs are gradually brought into play.

The gestation period for the mouse embryo is 19–20 days, depending on the strain. The timing of the different stages is shown in Table 4, which is based on the development of F_1 hybrids between C57BL/6 females and CBA males; for some inbred strains, such as C3H, the process is somewhat slower. Our current knowl-edge of cell lineages in the mouse embryo is summarized in Figure 2.

Although embryonic development starts with fertilization, both the egg and the sperm are themselves products of complex maturational processes initiated when the primordial germ cells enter the genital ridges. This summary of mouse development therefore begins with a description of the origin and growth of the germ cells, which are among the most fascinating cells in the whole organism.

The Origin of Germ Cells and Their Migration to the Genital Ridges

Primordial germ cells differentiate from cells of the primitive ectoderm and are first clearly identified in the 8- to 8.5-day postcoitum (p.c.) embryo at the base of the allantois (see Fig. 15E,F). They are distinguished both by their large, rounded shape and high levels of alkaline phosphatase activity. This activity can be revealed by histochemical staining and enables the germ cells to be traced during their

Table 4 Development of the Mouse Embryo

Stage[a]	Age (days p.c.)[b]	Features	System				
			Extraembryonic	Circulation	Intestinal tract	Nervous/sensory	Urogenital
1	0–1	one-cell egg					
2	1	two-cell egg					
3	2	morula, 4–16 cells					
4	3	morula-blastocyst	trophectoderm formed				
5	4	free blastocyst without zona					
6	4.5	implanting blastocyst	primitive endoderm				
7	5	egg cylinder					
8	6	proamniotic cavity in primitive ectoderm	Reichert's membrane forming	ectoplacental cone fills with maternal blood			
9	6.5	embryonic axis determined					
10	7	early–mid-primitive streak	amnion forming				
11	7.5	late primitive streak	allantois appearing	blood islands in visceral yolk sac	foregut pocket	neural plate	

Plate	Day	Somites	Circulation	Heart / aortic arches	Gut	Nervous system	Urogenital
12	8	1–7 somites	allantois contacts chorion	first aortic arch	hindgut pocket	neural folds, otic placode	germ cells near base of allantois
13	8.5	8–12 somites; turning of embryo		paired heart primordia fusing anteriorly	thyroid rudiment, second pharyngeal pouch, hepatic diverticulum	neural folds close at level of somites 4–5	pronephros
14	9	13–20 somites	blood circulates in visceral yolk sac	heart begins to beat, three paired aortic arches	oral plate ruptures	anterior neuropore closes, olfactory placode	pronephric duct still solid
15	9.5	21–29 somites; forelimb bud at level of somites 8–12		common ventricle and atrium, dorsal aortae fused	lung primordia, pancreas evagination, vitelline duct closed	posterior neuropore closes, otic vesicle	
16	10	30–34 somites; hindlimb bud at level of somites 23–28			primary bronchi	lens placode	Wolffian ducts contact cloaca in older specimens
17	10.5	35–39 somites; tail rudiment		sixth aortic arch	umbilical loop, cloacal membrane	deep lens pit	mesonephric tubules
18	11	40–44 somites		spleen primordium		lens vesicle closing, rims of olfactory placode fusing	distinct genital ridge

Table 4 (*Continued*)

Stage[a]	Age (days p.c.)[b]	Features	Extraembryonic	Circulation	System		
					Intestinal tract	Nervous/sensory	Urogenital
19	11.5	6- to 7-mm length, forefoot plate		partitioned atrium, unpaired ventricle	bucconasal membrane	lens vesicle detached	ureteric buds
20	12	7- to 9-mm, hindfoot plate		partition of arterial trunk begins	tongue, thymus, and parathyroid primordium	pineal body evaginates	sexual differentiation of gonads in older specimens
21	13	9- to 10-mm, whisker rudiments		aortic and pulmonary trunks separated	palatine processes vertical, dental laminae	lens solid	cloaca sub-divided
22	14	11–12 mm		interventricular septum closed		ganglionic cells of retina	separate opening of ureter into urogenital sinus
23	15	12–14 mm		coronary vessels	palatine processes fused		
24	16	14–17 mm	Reichert's membrane breaks down		reposition of umbilical hernia	eyelids fusing	
25	17	17–20 mm			alveolar ducts of lung	ciliary body delineated	large central glomeruli in kidney

26	18	19.5–22.5		pancreatic islands of Langerhans	iris and ciliary body	solid cords of prostate cells
27	19	23–27 mm	birth			testis cords still solid

[a] Adapted from Thieler (1972, 1983) based on the development of C57BL/6 × CBA F₁ hybrids. Embryos of some inbred lines of mice may develop more slowly.

[b] In this manual we use the following convention for timing pregnancy and the age of embryos. Assuming that fertilization takes place around midnight on a 7 pm–5 am dark cycle, then at noon on the following day (i.e., the day on which the vaginal plug is found) the embryos are aged "half-day postcoitum" or "0.5 day p.c." According to this convention, the day on which the plug is found is day 0 of pregnancy. At noon on the next day the embryos are 1.5 day p.c., and so on. Note, however, that the embryos of one litter are not synchronized in their development.

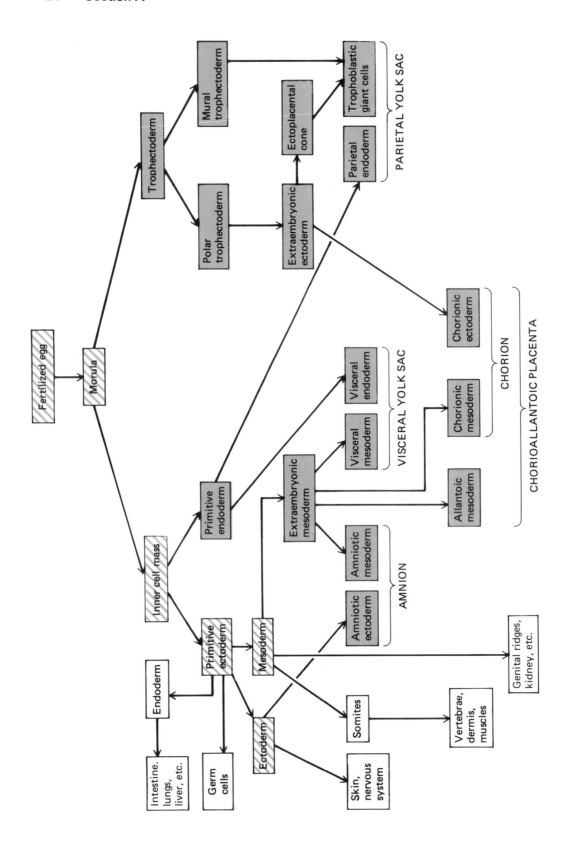

migration to the genital ridges (Clark and Eddy 1975; Eddy and Hahnel 1983) (Fig. 3). The conclusion that germ cells arise from the primitive ectoderm is based on experiments in which pieces of the 7- and 7.5-day p.c. egg cylinder (see Fig. 15E) are excised and cultured in vitro and then scored for the appearance of large alkaline phosphatase-positive cells (Snow 1981; Snow and Monk 1983). Only pieces excised from near the posterior end of the egg cylinder give rise to such cells.

As the germ cells move from the base of the allantois, they embark upon a complex migration, first into the endodermal epithelium of the hind gut, then into the mesentery, and finally into the genital ridges, which are bands of mesodermal tissue lying in the roof of the peritoneum. (For their isolation see Section C.) It is assumed that this migration involves some active movement of the germ cells, either along tracts of extracellular matrix material or in response to chemotactic substances released by the cells of the genital ridge.

En route to their final destination, the germ cells divide approximately once every 16 hours. It has been estimated that in the 8-day p.c. embryo there are between 10 and 100 primordial germ cells; this number increases to about 25,000 when the genital ridges are fully colonized at 13.5 days p.c. of development (Tam and Snow 1981). Several mutations are known that affect germ cell proliferation and cause sterility in homozygotes by severely reducing the number of germ cells in the genital ridges. For example, white spotting (W) and viable white spotting (W^v) reduce both the proliferation and migration of the germ cells from the base of the allantois (Mintz and Russell 1957). In contrast, in mice homozygous for Steel (Sl), the germ cells apparently migrate normally toward the genital ridges but fail to proliferate or be maintained there (McCoshen and McCallion 1975). Both mutations also affect neural crest cells and progenitors of the hematopoietic system, two other populations of cells that undergo migrations over long distances. In Sl/Sl but not in W/W mice, the pigment cells can be rescued by transplanting neural crest cells into the wild-type environment (Mayer 1973). However, it is not known whether Sl/Sl primordial germ cells can be rescued in the same way.

The Environment of the Genital Ridge Affects Germ Cell Development

Differences in the genital ridges of male and female embryos can first be detected at around 12.5 days p.c. (Fig. 4). At this time, the female germ cells enter meiosis in response to a stimulus provided by the somatic cells of the genital ridge (Monk and McLaren 1981; McLaren 1983). In contrast, male germ cells respond to the environment of the male genital ridge by entering mitotic arrest in the G_1 phase of the cell cycle and do not enter meiosis until later in development. Tissue culture techniques are now being developed (De Felici and McLaren 1983) that may make it possible to study the nature of the influence of the somatic cells of the genital ridge on germ cell differentiation in vitro (McLaren 1983). This effect of the local gonadal environment has a profound influence on germ cell differentiation in XX/

Figure 2 Summary of the lineages of tissues constituting the mouse embryo. (▨) All tissues that will give rise to the embryo proper and extraembryonic cells; (■) extraembryonic tissues; (□) tissues of the embryo proper. (Adapted from Gardner 1983.)

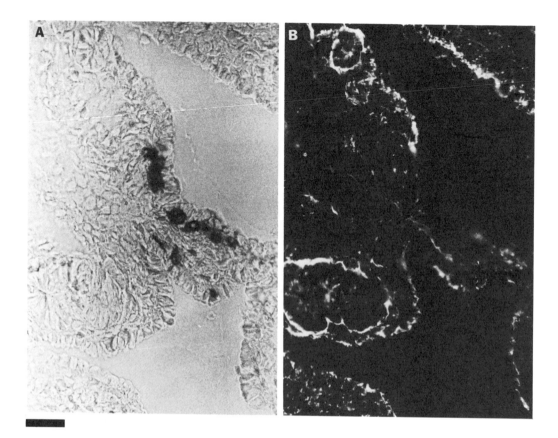

Figure 3   Germ cells in the 10.5-day mouse embryo. (*A*) Germ cells stained with alkaline phosphatase lying in the hind gut mesentery of a 10.5-day mouse embryo (bright-field illumination). (*B*) Immunofluorescence microscopy of an adjacent section stained to reveal laminin in the basement membranes of the mesentery and genital ridge. Scale bar, 20 μm. (Photographs provided by Dr. D. Stott, St. George's Hospital Medical School, London.)

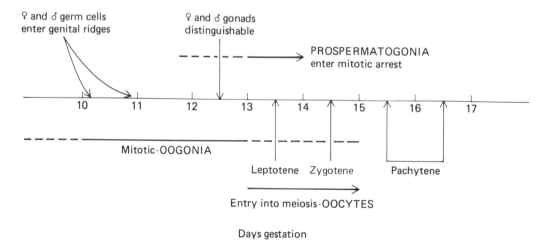

Figure 4   The timing of female and male germ cell development in the genital ridge and early gonad of the mouse. (Adapted from Monk and McLaren 1981.)

XY chimeras formed as a result of aggregating morulae (see Section C) or by injecting cells into the blastocyst (see Section D). XX germ cells in a predominantly XY local environment begin to develop in the male direction but do not form sperm, whereas XY cells in an XX environment can form oocytes (for review, see McLaren 1983).

Growth and Maturation of the Oocyte

By 5 days after birth all oocytes are in the diplotene stage of the prophase of the first meiotic division. They are therefore diploid but contain four times the haploid amount of DNA (4C). During the prolonged resting or dictyate stage, the paired homologous chromosomes are fully extended and transcription of oocyte (maternal) mRNA takes place. Studies on X chromosome activity have shown that only one X is active in XX primordial germ cells and 11.5 days p.c. oogonia, but that by 12.5 days p.c. both X chromosomes become active (Monk and McLaren 1981; McLaren 1983).

Each oocyte is contained within a follicle and is surrounded by multiple layers of follicle cells, which play various roles in oocyte growth and differentiation (Figs. 5 and 6). The follicle cells around the oocyte have numerous projections which form specialized junctions with the egg (Fig. 6). These junctional complexes involve gap junctions and allow metabolite transfer. They are maintained even as the follicle cells and oocyte are gradually separated by the deposition of the zona pellucida, a layer of extracellular material synthesized and deposited by the growing oocyte (Bleil and Wassarman 1980a,b; Greve and Wassarman 1985). The zona is composed of three major acidic sulfated glycoproteins (ZP1, M_r 200,000; ZP2, M_r 120,000; ZP3 M_r 83,000) and reaches a thickness of about 7 μm. Oocytes at various stages of maturity can be isolated from the ovary and cultured in vitro, with or without follicle cells (see Section C).

Apart from studies on the synthesis and processing of the zona glycoproteins, which together constitute about 10% of the total protein synthesis, relatively little is known about gene activity of growing oocytes. Several groups have carried out two-dimensional gel electrophoretic analysis of total [^{35}S]methionine-labeled proteins synthesized by maturing oocytes and unfertilized eggs (Van Blerkom 1981; Howlett and Bolton 1985). In addition, the synthesis of a number of specific proteins has been reported. For example, about 1.3% of the total protein synthesis of oocytes is devoted to tubulin (Schultz et al. 1979); oocytes also synthesize and secrete a protein of M_r 43,000 and of unknown function (Brinster et al. 1981a). Further evidence that the mouse oocyte is able to glycosylate and secrete proteins comes from studies on the fate of chick ovalbumin synthesized from exogenous mRNA injected into the cytoplasm (Paynton et al. 1983). Finally, it has been shown that several different repetitive sequences are undermethylated in mouse oocytes (Sanford et al. 1984).

Surprisingly, more than half of the primordial follicles present in the mouse ovary at birth degenerate before 3–5 weeks of age, but little is known about the hormonal and local factors controlling this loss (Faddy et al. 1983). The female mouse reaches sexual maturity at around 6 weeks of age, depending on the strain and environmental conditions. By this time each ovary contains approximately 10^4 oocytes at different stages of maturity.

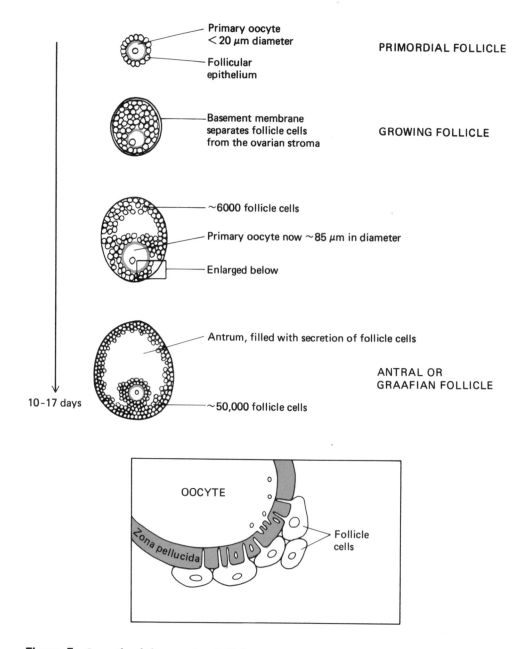

Figure 5   Growth of the ovarian follicle.

Ovulation

As the oocyte increases in size, it gradually acquires the competence to enter the final stages of meiosis in response to the correct hormonal stimulus. Ovulation requires the coordinated response of both the follicle cells and the oocyte, and under optimal laboratory conditions it occurs spontaneously once every 4 days. However, cycle length can be influenced by many environmental factors and can be induced artificially by hormone injection (superovulation, see Section C). In any

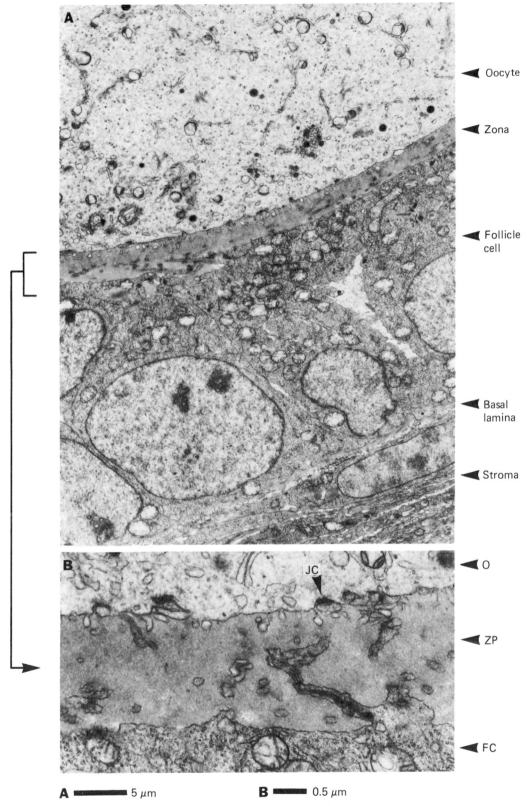

A ▬▬▬ 5 μm   B ▬▬ 0.5 μm

Figure 6 Relationship between oocyte and follicle cells. (O) Oocyte; (JC) junctional complex; (ZP) zona pellucida; (FC) follicle cell. Scale bars: *A*, 5 μm; *B*, 0.5 μm.

one natural cycle only a few follicles respond to an increase in the level of follicle-stimulating hormone (FSH), which is produced by the pituitary. The stimulated follicle cells break contact with the oocyte and increase their synthesis and secretion of high-molecular-weight proteoglycans and tissue plasminogen activator. At the same time the follicle accumulates fluid, swells, and moves toward the periphery of the ovary, ready for the final maturation and release of the oocyte. The mature, fluid-filled follicle units are known as antral or Graafian follicles after the scientist Regnier de Graaf who first described them in 1672. For an extensive review of the biosynthetic activity of follicle cells in vivo and in culture, see Hsueh et al. (1984).

Ovulation occurs in response to a surge in the level of luteinizing hormone (LH), also produced by the pituitary. After LH stimulation, the oocyte undergoes nuclear maturation (Fig. 7). The nucleus (which is also known as the germinal vesicle) loses its membrane and the chromosomes assemble on the spindle and move toward the periphery of the cell where the first meiotic division takes place. One set of homologous chromosomes, surrounded by a small amount of cytoplasm, is extruded as the first polar body. It is in this state that the oocyte is finally released from the follicle.

Each ovulated oocyte is surrounded by its zona and a mass of follicle cells (cumulus cells) with their associated proteoglycan. The eggs are swept into the open end, or infundibulum, of the oviduct by the action of the numerous cilia on the surface of the epithelium. Other cells in the epithelium are secretory, and at the time of ovulation the distal end of the tube becomes engorged and enlarged to form an ampulla where fertilization takes place. In a natural ovulation, 8–12 eggs are released (depending on the mouse strain) but the process is not synchronous and occurs over a period of 2–3 hours. After ovulation, the follicle cells remaining behind differentiate into steroid-secreting cells (luteinized granulosa cells) which help to maintain pregnancy. Counting the number of bright yellow corpora lutea near the surface of the ovary is a way of determining how many eggs were, in fact, released.

Fertilization

Approximately 58×10^6 sperm are released into the female reproductive tract per ejaculation. Some sperm reach the ampulla within 5 minutes, but they are not competent for fertilization for about 1 hour. This process of maturation is known as capacitation, but its mechanism is unknown. In order to reach the surface of the egg the sperm must penetrate first the cumulus mass and then the zona pellucida. In many mammals, the sperm binding sites in the zona are highly (but not absolutely) species specific and prevent the passage of sperm from other species. Sometime during the penetration of the cumulus or zona the sperm undergoes the acrosomal reaction. The acrosome (a secretory vacuole-like structure in the sperm head) fuses with the plasma membrane of the sperm head, releasing various hydrolytic enzymes. Unless the acrosomal reaction takes place, the sperm cannot fertilize the egg. In the mouse, the smallest zona protein, ZP3, has been implicated in triggering the acrosomal reaction and may serve as the sperm receptor (Bleil and Wassarman 1983).

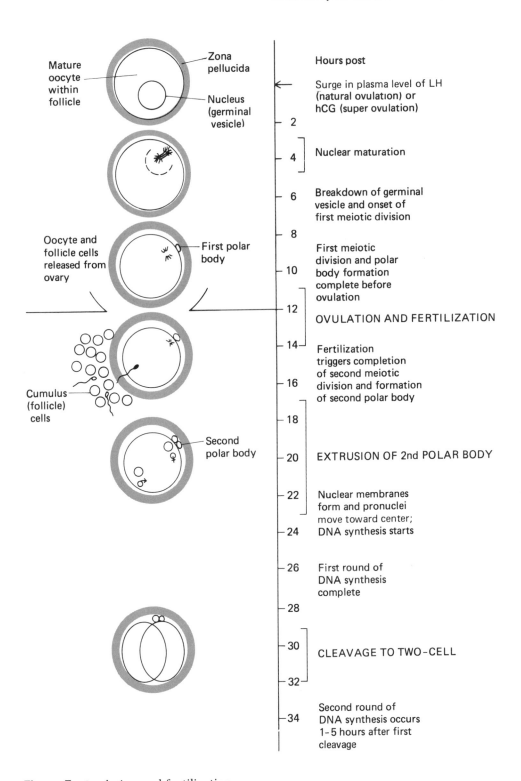

Figure 7 Ovulation and fertilization.

Fusion of the posterior part of the sperm head with the egg membrane triggers the cascade of reactions known as fertilization. One very early event is a change in the egg surface inhibiting the fusion of additional sperm. Another event is the Ca^{++}-dependent release (exocytosis) of the cortical granules positioned beneath the plasma membrane. This initiates the "zona reaction" which involves both cross-linking of the glycoproteins of the zona and modification of the ZP3 glycoprotein so that it no longer binds sperm or elicits the acrosomal reaction. These events also help to prevent polyspermy. During fertilization, the head, midpiece, and a large part of the tail of the sperm are all incorporated into the egg cytoplasm. The midpiece of the sperm contributes paternal centrioles and mitochondria to the zygote, but the latter are enormously diluted out by the mitochondria of the oocyte.

Fertilization triggers the second meiotic division and extrusion of the second polar body. Nuclear membranes, including nuclear lamin proteins, then form around the maternal and paternal chromosomes, and the haploid female and male pronuclei move toward the center of the egg. DNA replication takes place during this migration. The pronuclei do not fuse, but the membranes break down and the chromosomes assemble on the spindle; the first cleavage occurs soon after. Because of asynchrony in ovulation and fertilization, the first cleavage occurs over a number of hours in a population of naturally fertilized zygotes. More synchronous development can be obtained by in vitro fertilization (see Section C). Unfertilized eggs remain viable for about 12 hours and sperm for about 6 hours.

Some properties of the ovulated unfertilized mouse oocyte are given in Table 5 and the sequence of events from onset of nuclear maturation to blastocyst formation is illustrated in Table 6 and Figures 7, 8, and 9.

Table 5   Some Properties of the Ovulated, Unfertilized Mouse Oocyte

Diameter	85 μm
Volume	270 pl (volume of pronucleus 1 pl)
Protein	23 ng
Total DNA	8 pg
Mitochondrial DNA	2–3 pg (note that much of the DNA of the unfertilized egg is mitochondrial)
Number of mitochondria	10^5
Genomic DNA (haploid number of chromosomes but diploid [2C] amount of DNA)	6 pg
Ribosomal RNA	0.2–0.4 ng
Poly (A)	0.7 pg (120–200 nucleotides long)
Poly (A) + RNA	exact amount not determined
tRNA	0.14 ng

Table 6 Timing of Events During the First Cell Cycle of
Embryos Derived from (C57BL ♀ × CBA ♂) F₁ ♀ × CFLP
(Outbred) ♂ Following In Vitro Fertilization

Event[a]	Hours postinsemination
Extrusion of second polar body	2–5
Formation of ♂ pronucleus	4–7
Formation of ♀ pronucleus	6–9
DNA replication	11–18
Cleavage	17–20

Information supplied by S. Howlett, Department of Anatomy,
University of Cambridge, U.K.

[a]For morphology, see Fig. 4.

Cytoskeletal Organization of the Egg before and after Fertilization

Throughout the cytoplasm of the oocyte there is a complex matrix of cytoskeletal elements, including actin, tubulin, and certain cytokeratins (Lehtonen et al. 1983a; Maro et al. 1984; Schatten et al. 1985) (Fig. 10). The different systems must help to coordinate events at the cell surface with changes in the pronuclei as they migrate toward the center of the egg. This migration is inhibited by both cytochalasin B (which inhibits actin polymerization) and colcemid (inhibits tubulin polymerization), and both inhibitors are required together in nuclear transfer experiments to enable the nuclei to be withdrawn into a karyoplast (see Section D). The earliest developmental changes in actin organization in the egg are seen at fertilization (Maro et al. 1984). In the ovulated oocyte the plasma membrane above the meiotic spindle is devoid of concanavalin A (Con A) binding sites and microvilli, and is underlaid by an actin-rich subcortical layer. Fertilization results in the formation of a second Con A free zone, the fertilization cone, which is around the site of sperm entry. The plasma membrane in this region is also underlaid by an actin-rich layer. As the pronuclei move toward the center of the egg, the distribution of actin filaments becomes more uniform and the Con A-free regions disappear.

Parthenogenesis and the Need for Both Maternal and Paternal Genomes for Complete Development

Reviews of the experimental production of parthenogenetic embryos and their contribution to developmental biology are to be found in Kaufman (1981, 1983a) and Whittingham (1980b). Parthenogenetic activation of unfertilized eggs can be elicited by exposing them to a variety of agents, including alcohol (see Section C and Fig. 37), hyaluronidase, the Ca^{++} ionophore A23187, Ca^{++}/Mg^{++}-free medium, heat/cold shock, or anesthetics. In addition, about 10% of the oocytes of the LT/Sv strain of mice undergo spontaneous activation with high frequency, either in the oviduct or in the ovary. Those embryos that implant develop to the egg cylinder stage (7 days p.c.) and then become disorganized and die, while those remaining in the ovary give rise to teratomas.

The genotype of the parthenogenetic embryo can vary depending on the ex-

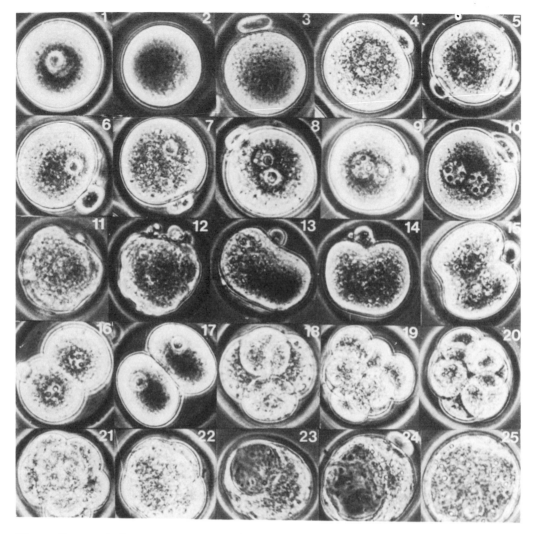

Figure 8 Morphology of preimplantation mouse development. (*1*) Preovulatory oocyte with germinal vesicle intact. (*2*) Preovulatory oocyte showing breakdown of germinal vesicle (GVBD) (2.5–4.5 hr post-hCG). (*3* and *4*) Extrusion of first polar body (~10 hr post-hCG) followed by ovulation (~11–13.5 hr post-hCG) and fertilization. (*5*) Resumption of meiosis by female set of chromosomes and extrusion of second polar body (occurs over the period 17–23 hr post-hCG). (*5* and *6*) Decondensation of sperm nucleus and formation of male pronucleus. (*7*) Formation of nuclear membrane around haploid set of female chromosomes to form female pronucleus which is subcortical, near the second polar body, and smaller than the male pronucleus (process complete in the majority (75%) of the embryos by ~26 hr post-hCG). (*8*) Migration of pronuclei to center of egg. (*9* and *10*) Formation of visible nucleoli within both pronuclei. DNA replication (complete in the majority of embryos by ~28 hr post hCG). (*11* and *12*) Breakdown of pronuclear membranes and disappearance of visible nucleoli. Ruffling of embryo surface indicating reorganization of the cytoskeleton preparatory to cleavage (observed from ~27 hr post-hCG until cleavage is completed). (*13*) Elongation of embryo. (*14* and *15*) Formation of "waist." (*16*) Newly formed two-cell embryos with visible nucleoli (the majority of embryos have cleaved by ~32 hr post-hCG). (*17*) Later-stage two-cell embryo with nuclei visible. (*18–25*) Later stages of preimplantation development. (*18*) Four-cell embryo. (*19–20*) Six to eight-cell embryo. (*21*) Compacting eight-cell embryo. (*22*) Compacted eight- to 16-cell embryo. (*23* and *24*) Early blastocysts. (*25*) Fully expanded blastocyst. (All timings are given in hours after injection of hCG. Information and figure provided by Dr. H. Pratt, Department of Anatomy, University of Cambridge.)

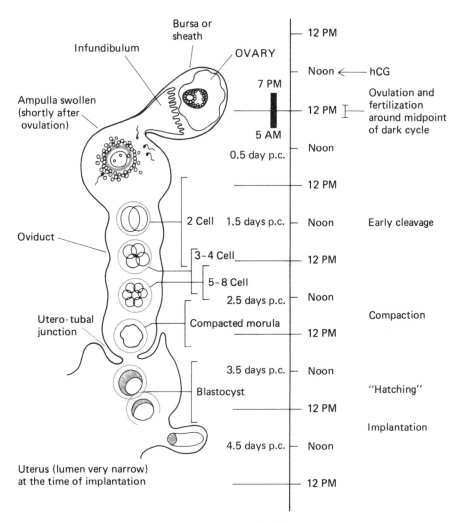

Figure 9 Summary of preimplantation development.

perimental conditions and, in particular, on the postovulatory age of the activated oocyte. The most important factor may be the state and orientation of the cytoskeletal elements in the egg when activation takes place. Among the possible genotypes arising from the parthenogenetic activation of an oocyte from an F_1 (heterozygous) female are (Fig. 11):

1. Uniform haploid (second polar body successfully extruded).

2. Mosaic haploid (second polar body behaves as a normal blastomere).

3. Heterozygous diploid (results from suppression of second polar body formation or from fusion of the pronucleus and the second polar body). The heterozygosity in these parthenogenones is the result of recombination during meiosis.

4. Homozygous diploid (results from diploidization of the female pronucleus).

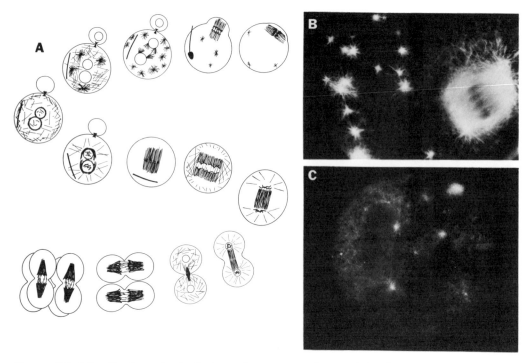

Figure 10 Distribution of tubulin in early mouse embryos. (*A*) Schematic representation of tubulin distribution from unfertilized egg through to eight-cell stage. (*B*) Unfertilized egg showing cytoplasmic asters. (*C*) Pronuclear-stage fertilized egg. (Photograph provided by Dr. Gerald Schatten, Department of Biological Science, The Florida State University.)

Note also that up to 20% of eggs activated by alcohol may be aneuploid as a result of nondisjunction (Kaufman 1982, 1983b). Most parthenogenones, and particularly the uniform haploids, die before the blastocyst stage. A minority continue to develop past implantation, for example up to the egg cylinder stage (LT/Sv and mosaic haploid parthenogenones) and the early limb bud stage (heterozygous diploids; Kaufman et al. 1977). However, no normal development to term has been obtained from any class. The reason for this failure is not yet understood, but cell lethality is clearly not a factor, for several reasons. First, when transplanted to ectopic sites, both LT/Sv and other parthenogenones can generate teratomas containing a whole variety of normally differentiated but disorganized tissues. Second, they can be rescued by chimera formation with normal embryos (Stevens et al. 1977; Surani et al. 1977). In addition, pluripotent cell lines have recently been established from haploid parthenogenetic blastocysts cultured in vitro (Kaufman et al. 1983). In these lines and in teratomas derived from parthenogenones, the cells soon diploidize.

The failure of parthenogenetic embryos to develop to term points to a defect in some process normally required for the coordinated growth and development of the embryo as a whole, rather than cell viability per se. One possibility is that inbred strains of mice carry recessive lethal mutations in genes controlling tissue

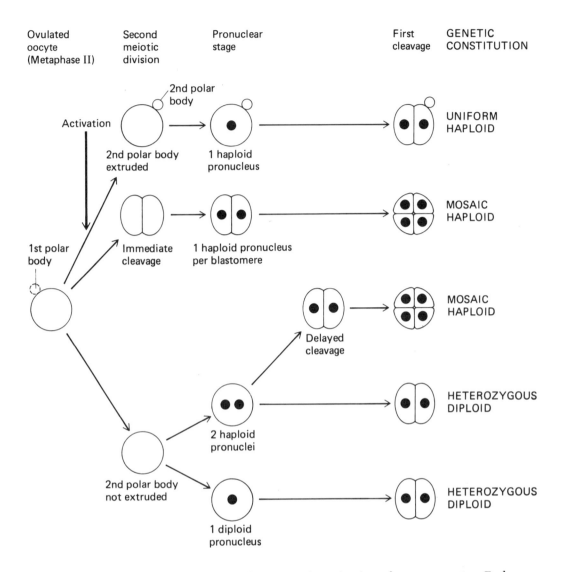

Figure 11 Possible products of parthenogenetic activation of mouse oocytes. Embryos of different genotype are produced depending on whether a second polar body is extruded or not, and on the timing and nature of the first cell division.

organization or pattern formation. Making the simple assumption that such recessive mutations would show complete penetrance in homozygotes, this explanation is unlikely because it predicts that 25% of the offspring of inbred strains of mice would die in utero. Although prenatal mortality can be quite high in some inbred strains (see references in Table 1), it does not always reach this level. Another possibility is that some nongenetic contribution of the sperm, or some cytoplasmic event triggered only by fertilization, is required for normal embryonic development. This latter view was supported initially by the results of Hoppe and Illmensee (1981). They transplanted nuclei from inner cell masses of parthenogenetic LT/Sv blastocysts into C57BL fertilized eggs and then removed both host pronuclei. A small proportion of the injected eggs developed normally (Hoppe and Illmensee 1981). In earlier experiments they had apparently obtained some viable offspring from fertilized eggs after removing one of the pronuclei and allowing the other to diploidize in the presence of cytochalasin (Hoppe and Illmensee 1977). However, similar experiments to produce homozygous diploid uniparental mice have not been successful in other laboratories (for review; see Markert 1982; Surani and Barton 1983).

More recent experiments in two independent laboratories have provided evidence in support of a third hypothesis: that normal development only occurs if one set of chromosomes has gone through an "imprinting" stage in the male germ line. In other words, the cytoplasm of parthenogenetically activated eggs is fully competent to support development to term provided that both male and female pronuclei contribute to the embryonic genome (McGrath and Solter 1984b; Surani et al. 1984). Both laboratories have used the nuclear transplantation technique devised by McGrath and Solter (1983a,b) and described in Section D. Surani and his colleagues showed that if a male pronucleus is transferred to a parthenogenetically activated haploid egg the resulting embryo has a high chance of developing to term. If a female pronucleus is transferred, however, the embryos will not develop beyond about 10 days, and there is particularly poor growth of the extraembryonic tissues. The need for both maternal and paternal genomes for complete development has also been demonstrated by transplantation of pronuclei between fertilized eggs (McGrath and Solter 1984b). One explanation of these results is that there is a high frequency of errors in X chromosome inactivation in the extraembryonic tissues when both X chromosomes are derived only from the female. Normally, there is preferential inactivation of the paternal X chromosome in extraembryonic tissues as will be discussed later (see below). The molecular basis of the imprinting of the paternal genome is not yet known.

Early Cleavage: One-cell Embryo to Eight-cell Uncompacted Morula

In spite of the small size of cleavage-stage mouse embryos, a considerable amount of information is now available about changes in the pattern of RNA and protein synthesis during preimplantation development. Some progress has also been made in correlating these changes with events such as fertilization, DNA replication, cell division, and commitment of blastomeres to different developmental fates. This information has been summarized in two recent reviews (Johnson 1981; Pratt et al. 1983).

As described in these reviews, two-dimensional SDS-polyacrylamide gel electrophoresis (PAGE) has revealed a number of changes in the pattern of [^{35}S]methionine-labeled proteins synthesized by early-cleavage embryos. It has been difficult to time these changes precisely, particularly in relation to other cellular events (for example, DNA synthesis and cell division) because of asynchrony both within a population of normally fertilized embryos and between blastomeres of individual embryos. This asynchrony can be reduced by in vitro fertilization, by picking out embryos after cleavage to two-cells, and by dissociating and recombining groups of blastomeres at specific stages of the cell cycle before labeling. Although prolonged exposure to [^{35}S]methionine may have deleterious effects due to radiation damage which cannot be repaired (MacQueen 1979), exposure of a few hours apparently has no effect on subsequent development (Van Blerkom 1981).

Several processes may be responsible for the changes in the pattern of protein synthesis seen after fertilization.

1. Increased turnover rates of some proteins made on stable maternal mRNAs. Recently evidence for such mechanisms has been found in mammalian eggs (Howlett and Bolton 1985.)

2. Posttranslational modification of proteins synthesized on either maternal or embryonic RNA. There is some evidence for modification of proteins by phosphorylation, glycosylation, or proteolytic cleavage (Van Blerkom 1981; Cascio and Wassarman 1982; Pratt et al. 1983).

3. Selective use of subspecies of maternal mRNAs. There is now clear evidence from the study of carefully timed embryos and from comparison of in vitro and in vivo translation products that some mRNA species are utilized or suppressed selectively.

4. Specific degradation of maternal mRNAs carried over from the oocyte. At the two-cell stage there is a sharp fall in the level of total and poly(A)$^+$ RNA (Clegg and Piko 1983) in the translation of globin mRNA injected into the zygote (Brinster et al. 1980) and in the translation of proteins thought to be coded for by maternal RNA (for review, see Johnson 1981; Pratt et al. 1983). In a particularly clear series of experiments using Northern hybridization to recombinant DNA probes, Giebelhaus et al. (1983) showed a marked loss at the two-cell stage of the pools of actin and histone H3 mRNA present in the egg. Subsequently, the level of these mRNAs in the embryo began to rise after the four-cell stage, when transcription from the embryonic genome is underway.

5. Synthesis of proteins on mRNAs transcribed de novo from the embryonic genome. Experiments in which eggs are incubated in the presence of the RNA polymerase inhibitor α-amantin have shown that new RNA synthesis is required for development beyond the two-cell stage and for the synthesis of many new proteins (Flach et al. 1982). Transcription of both ribosomal and poly(A)$^+$ RNA, which is apparently shut off at the time of germinal vesicle breakdown, is thought to resume at a low level around the mid-two-cell stage. Timing of the onset of synthesis of paternally coded proteins has come from studies on the expression of genetic enzyme variants or antigens, and the synthesis of a few specific proteins has been followed by metabolic labeling and immunoprecipitation (Table 7).

In summary, up to the mid-two-cell stage (27 hr postfertilization) the embryo appears to rely largely on protein and RNA synthesized during oogenesis. By the mid-two-cell stage, many embryonic genes are switched on. Coincidentally much of the maternally inherited mRNA appears to be degraded rapidly. However, maternally coded proteins can persist beyond this time.

Table 7 Some Specific Proteins Made by Mouse Embryos from the Two-cell to Compacted Morula Stage

Protein	Stage first detected	Methods used	Reference
β-Microglobulin	two-cell	electrophoretic variant, metabolic labeling, and immunoprecipitation	Sawicki et al. (1981)
Intracisternal A-type particle-associated antigens	zygote → eight-cell peaks at two-cell	variety of techniques used	Huang and Calarco (1981)
Laminin	fertilized and unfertilized egg, not at two-cell stage, then four-cell stage onwards	metabolic labeling and immunoprecipitation chains not coordinately expressed	Cooper and MacQueen (1983)
β-Glucuronidase	four-cell	genetic enzyme variant	Wudl and Chapman (1976)
Non-H-2 alloantigen	six- to eight-cell	immunological detection	cited in Johnson (1981)
HY antigen	eight-cell	immunological detection	Krco and Goldberg (1976)
6-Phosphogluconate dehydrogenase	eight-cell	genetic enzyme variant	cited in Johnson (1981)
Glucose phosphate isomerase (GPI)	eight-cell	genetic enzyme variant	cited in Johnson (1981)
HPRT	eight-cell	deduced from enzyme activity levels in XY versus XX embryos	Epstein et al. (1978)
90K Heat-shock protein	morula and eight-cell stage	two-dimensional gel electrophoresis	quoted in Bensaude and Morange (1983)
68K and 70K heat-shock proteins (or cognate proteins)	G_1 of 2-cell stage	earliest expressed protein	Bensuade et al. (1983)

Compaction and the Formation of the Blastocyst: The First Differentiation Events

Up to the early-eight-cell stage there is good evidence that the blastomeres of the mouse embryo are equipotent. Single blastomeres from two-and four-cell morulae can each give rise to a mouse. Early eight-cell-stage blastomeres cannot generate a mot   by themselves, but when recombined with genetically marked morulae they (ι give rise to a wide range of different tissues in chimeric offspring (Kelly 1977). As cleavage proceeds to the 16-cell stage, however, there is a gradual restriction in the developmental potency of the cells, resulting in the generation of two distinct lineages: the trophoectoderm (TE) and the inner cell mass (ICM). This differentiation process starts with compaction, when the blastomeres flatten and increase their contact with each other and develop distinct apical and basal membrane and cytoplasmic domains (polarization). It ends with the formation of a fully expanded blastocyst consisting of a hollow vesicle of trophoectoderm surrounding a fluid-filled cavity (the blastocoel) and a small group of ICM cells. The trophoectoderm has all the features of a true epithelium with apical junctional complexes forming a complete permeability seal against the outside environment.

Some of the cellular changes associated with compaction are listed in Table 8. It should be stressed that they do not occur synchronously within all the cells of one embryo. Likewise, the cell cycles are not synchronized (Graham and Deussen 1978). The changes associated with compaction clearly point to alterations in both the surface properties of the cells and the organization of the cytoskeleton. The molecular basis of these changes, the signal(s) eliciting them, and their relationship to each other and to the cell cycle are areas of active research.

Changes in Cell Adhesiveness with Compaction

From the point of view of cell adhesiveness, there is good evidence that compacting embryos do not synthesize the extracellular matrix proteins fibronectin, or collagens (I through IV) (Wartiovaara et al. 1979; Leivo et al. 1980). Laminin is synthesized by morulae (Cooper and MacQueen 1983 and Table 8) and can be localized histochemically between the cells of the compacted morula (Leivo et al. 1980; Wu et al. 1983) (Fig. 12). However, polyvalent rabbit antiserum directed against the native protein does not inhibit or reverse compaction (A.R. Cooper and H. MacQueen, unpubl.) and neither does the addition of exogenous purified laminin (J.C. Chisholm, M.H. Johnson, and B.L.M. Hogan, unpubl.). In contrast, compaction is inhibited by polyvalent rabbit antibodies (either whole serum or IgG Fab fragments) raised against F9 embryonal carcinoma cells (Kemler et al. 1977). This decompaction effect appears to be mediated at least in part via binding to a surface glycoprotein known as uvomorulin (Hyafil et al. 1980, 1981; Peyrieras et al. 1983). The M_r of this component is approximately 123,000 but in the presence of Ca^{++} it is cleaved by trypsin to give a M_r 84,000 glycoprotein fragment. In the absence of Ca^{++} this fragment is further cleaved by trypsin to low-molecular-weight material. These results point to a Ca^{++}-dependent configurational change in uvomorulin, which may be reflected in the Ca^{++} dependence of compaction. Recent evidence has shown that uvomorulin is the same as the glycoprotein "cadherin" identified on F9 as cells being involved in Ca^{++}-dependent intercellular adhesion and compaction (Yoshida and Takeichi 1982; Shirayoshi et al. 1983; Yoshida-Noro et al. 1984).

Table 8 Changes Occurring in Blastomeres during Compaction

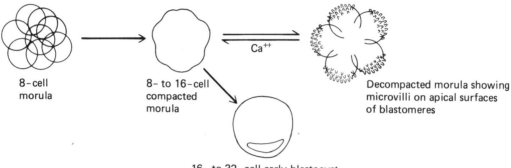

16- to 32–cell early blastocyst

1. Increased Ca^{++}-dependent adhesiveness, both to each other and to lectin-coated beads.

2. Increased spreading on adhesive surfaces, using lamellipodia-like cell processes.

3. Ability to express contact-induced cell polarization as shown by regionalization of membrane and cytoplasmic domains (microvilli, lectin binding sites, and intracellular organelles).

4. Establishment of gap junction-mediated intercellular communication (ionic coupling and dye transfer) between all cells of the morula unit.

5. Gradual development of apical, zonular tight junctions between outside cells, generating an impermeable outer epithelial layer.

Agents that inhibit features of compaction include cytochalasin B, tunicamycin, low Ca^{++}, rabbit anti-F9 embryonal carcinoma serum or Fab fragments, and some monoclonal antibodies against uvomorulin all tend to prevent or reverse the cell-spreading effects but not polarization.

Uvomorulin may also be identical to the cell adhesion molecule CAM 120/80 described by Damsky et al. (1983) and to L-CAM, a sialated surface glycoprotein involved in the Ca^{++}-dependent intercellular adhesion of liver and other tissues. It is important to stress that synthesis of uvomorulin is not unique to compacting morulae. It can be detected on the cell surface prior to compaction (Hyafil et al. 1981) and it is made by embryonal carcinoma cells and by cells of later-stage embryos, as well as by many epithelial tissues in adult animals (Damsky et al. 1983).

Cell Polarization with Compaction

One of the essential features of compaction is the polarization of the blastomeres, so that they show distinct apical and basolateral membrane domains. These domains are clearly seen by scanning electron microscopy of compacted embryos

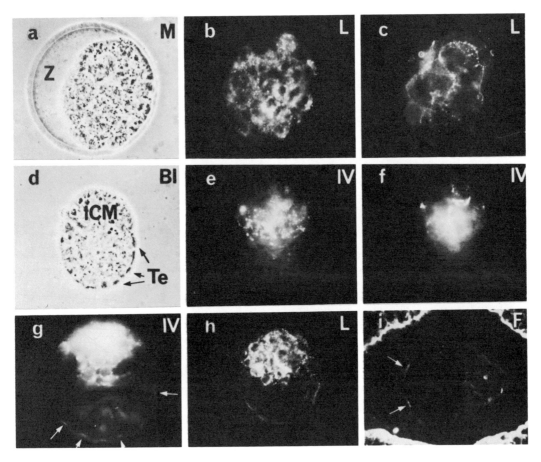

Figure 12 Preimplantation embryos stained for extracellular matrix proteins. (*a*) Phase contrast, early 16- to 32-cell morula. (Z) Zona pellucida. (*b*) As above, fixed with paraformaldehyde and treated with nonionic detergent to reveal both extracellular and cytoplasmic staining with laminin antibodies. (*c*) As above, but without detergent treatment, so that staining reveals only extracellular laminin. (*d*) Phase-contrast 3-day blastocyst. (ICM) Inner cell mass; (Te) trophectoderm. (*e* and *f*) As above, after fixation and detergent and staining with type-IV collagen antibodies. Both cytoplasmic and extracellular staining is seen, in two focal planes. (*g*) As above but a 4-day blastocyst. Type-IV collagen is now present on the inner surface of the trophectoderm (arrows). (*h*) A 4-day blastocyst as above, but stained with laminin antibodies. (*i*) Section of 4.5-day implanting blastocyst, stained with fibronectin antibodies. (This figure was provided by Dr. I. Leivo, University of Helsinki.)

which have been dissociated by incubation in the absence of calcium (Reeve and Ziomek 1981); the outer poles of the cells have numerous microvilli, while the inner surfaces are smooth. Cytoplasm organelles also appear to be polarized after compaction, with nuclei taking up a basal position. The onset of polarization can be followed in vitro by incubating pairs of isolated precompaction blastomeres. During culture the microvillous surfaces and Con A binding sites always develop at the poles opposite the points of cell–cell contact (Ziomek and Johnson 1980; Johnson and Ziomek 1981). An important question under investigation is whether this redistribution of plasma membrane domains precedes, or results from, a reorganization of cytoskeletal elements. Recent work in the laboratory of Dr. Martin Johnson, University of Cambridge, supports the idea that the cytoplasmic polarization that develops at the eight-cell-stage in vivo is dependent upon a functional cytoskeletal system. The rules that emerge about polarization of mouse embryos at the morula stage may apply more generally to the differentiation of epithelial tissues from nonpolarized precursor cells at later stages of development.

Segregation of the Trophectoderm and Inner Cell Mass Cell Lineages

As outlined above and in Table 8, compaction is associated with cellular polarization. This property forms the basis of a polarization hypothesis to account for the differentiation of the two distinct cell lineages of the blastocyst, the trophoectoderm, and the ICM. Cleavage planes through compacted morula cells horizontal to the polarized axis will generate basal or inside cells, and apical or outside cells, each inheriting different membrane and cytoplasmic molecules (for example plasma membrane glycoproteins, receptors, cytoskeletal organizing centers, etc.). These inherited molecules are thought to be responsible for initiating differences in the developmental potentials of the inner and outer cells. According to this hypothesis, differentiation is the result of cellular polarization elicited early in compaction. According to an alternative inside/outside microenvironment hypothesis, differentiation does not occur until after a network of tight junctions between the outer cells has formed. This generates distinct inside and outside microenvironments to which the cells respond, inside cells by becoming ICM and outside cells by becoming trophoectoderm.

A full description of compaction and a discussion of the polarization and microenvironment theories can be found in Johnson et al. (1981), Johnson and Ziomek (1981), Pratt et al. (1981), and Gardner (1983).

Restriction in the Developmental Potential of Nuclei

Nuclear transplantation experiments of Illmensee and Hoppe (1981) suggested that irreversible modification of the genome did not occur until the differentiation of the trophectoderm (TE). This was based on experiments in which nuclei transplanted from TE cells of 4-day blastocysts into enucleated fertilized host eggs failed to support development beyond early cleavage. On the other hand, in a small percentage of experiments, nuclei transplanted from ICM cells did apparently support development to viable and fertile offspring.

More recent experiments by McGrath and Solter (1984a,b,c), however, present a very different picture. They find that even nuclei transplanted from cleavage-stage embryos into enucleated eggs are unable to support development beyond the blastocyst stage. They argue for a close synchrony between nuclear events, cell division, and the redistribution of cytoplasmic molecules, which is disrupted by transplantation between stages. It is as though fertilization triggers the running down of clocks in both the nucleus and cytoplasm, and these clocks must be synchronized for correct development; the greater the time difference (e.g., eight-cell nucleus into one-cell cytoplasm), the poorer the development observed.

Implantation

The fully expanded blastocyst contains about 64 cells, of which about 20 are in the ICM (Fig. 13). The ICM can be separated from the trophectoderm by immuno-surgery (Solter and Knowles 1975; see Section C) or by microsurgery. During the 5th day of development the blastocyst hatches from the zona and is ready for implantation. Hatching is brought about by a trypsin-like enzyme (strypsin) synthesized by cells in the mural trophoblast (Wassarman et al. 1984) and by the expansion and contraction of the blastocyst. Around the time of hatching, the walls of the uterus become tightly apposed so that the lumen is closed, and changes in the uterine epithelium produce a surface conducive to blastocyst attachment. There are no preformed crypts into which the blastocysts collect; rather each embryo is thought to induce the formation of a crypt in the process of implantation, and the spacing of embryos along the uterus is due to the peristaltic movement of the uterus.

During the process of implantation, the mural trophoblast cells (see below) first invade the uterine epithelium and its underlying basal lamina and then penetrate the uterine stroma (also known as the endometrium) (Fig. 14). There is evidence for some local death of uterine epithelial cells around the site of implantation and active phagocytosis of these cells by the trophoblast (Enders et al. 1981). Blastocysts cultured in vitro (see Section C) will attach to tissue culture plastic surfaces and to layers of cells and cell-free extracellular matrix. In vitro culture on an extracellular matrix has recently been used as a model to investigate the molecular basis of trophoblast invasiveness (Glass et al. 1983).

If the pregnant female is ovariectomized, blastocysts will not implant in vivo but will remain viable or "delayed" in the uterus (for description of this technique, see Section C). This is presumably because the uterine epithelium does not present a suitable substratum for attachment, and subsequent treatment of the female with hormones leads to rapid implantation. Delayed blastocysts contain more ICM cells than normal, but these do not progress in their differentiation beyond the formation of primitive endoderm (for references, see Evans and Kaufman 1981). The mesenchymal stromal cells of the endometrium and smooth muscle in the myometrium respond to implantation of the embryo (or to inert substances such as oil or hair) in a number of interesting ways. This is known as the decidual response and is dependent upon progesterone. Increased mitosis is seen, and the cells, which previously had few intercellular junctions, establish numerous tight junctional complexes with their neighbors. Blood vessels in the deciduum lose their basement membranes and break down to give blood sinuses (for a review of the decidual response, see Finn 1971).

Trophectoderm and Its Derivatives

As described above, an essential feature of the differentiation of the trophectoderm of the early blastocyst is the organization of the cells into a typical epithelium. The cells have apical junctional complexes and distinct apical and basal membrane domains. The junctional complexes involve extensive desmosomes (Fig. 13), and associated with these are large numbers of intermediate filament bundles (Jackson et al. 1980).

During postimplantation development, the trophectoderm does not remain as

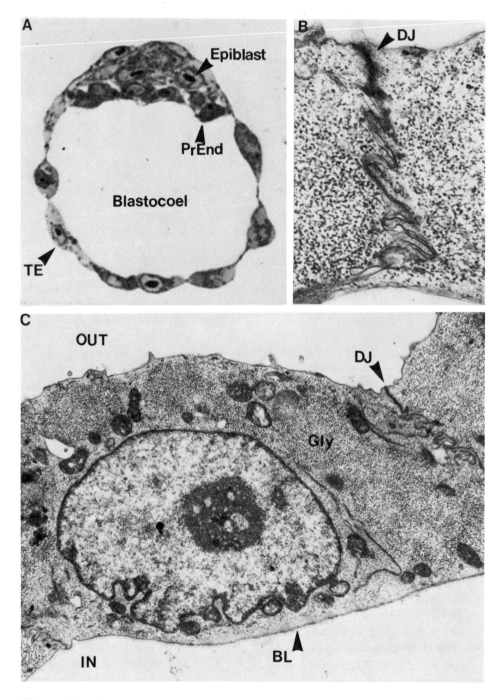

Figure 13 The mouse blastocyst at ~4.5 days of development. (*A*) Section showing outer epithelial vesicle of trophectoderm (TE) surrounding the blastocoel cavity and the epiblast and primitive endoderm (PrEnd). (*B*) Junction between two trophectoderm cells showing desmosomal junction (DJ) and interdigitation of the plasma membranes. (*C*) Trophectodermal cells showing abdundant cytoplasmic glycogen, desmosomal junction, and thin basal lamina (BL) on the internal surface.

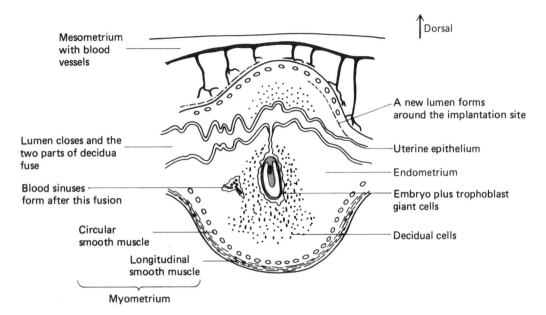

Figure 14 Relationship between the embryo and the deciduum about 2 days after implantation (6.5 days p.c.). The diagram represents a longitudinal section through the uterus, and shows the invariant orientation of the egg cylinder with respect to the dorsal–ventral axis of the mother. For a full discussion of how the mouse embryo may establish its dorso–ventral, anterior–posterior, and left–right axes of symmetry, see Smith (1985).

a simple epithelium but becomes regionally specialized with respect to morphology and growth potential (for review, see Gardner 1983). One subpopulation, the mural trophectoderm, is derived from the cells that surround the blastocoel cavity but are not in contact with the ICM. These cells cease division and become large and polyploid; they can contain up to 1000 times the haploid amount of DNA, and 160 chromosomes have been counted in one spread (Ilgren 1981). These are the so-called primary trophoblastic giant cells. In contrast, the trophectoderm cells in close proximity to the ICM and its derivatives remain diploid and continue to proliferate rapidly.

After implantation this population of so-called polar trophectoderm spreads in several directions. First, some cells migrate around the embryo, replacing the primary mural trophoblastic giant cells and themselves becoming polyploid. Second, a finger-like projection of polar trophectoderm penetrates down into the blastocoel cavity forming the extraembryonic ectoderm of the egg cylinder and pushing the ICM derivatives ahead of itself. This projection develops a central cavity and becomes epithelial. After the formation of the extraembryonic mesoderm the extraembryonic ectoderm moves back toward the placenta where it forms the chorion. Finally, some trophectoderm cells continue to penetrate into the endometrium forming the bulk of the placenta. Some of these cells, and cells of the chorion, also become polyploid (secondary giant cells).

Proliferation of the trophectoderm appears to be controlled by its proximity to ICM derivatives; in the absence of ICM derivatives TE cells do not proliferate but instead become giant. This has obvious advantages in preventing continued growth of trophectoderm if the embryo dies in utero.

The Second Round of Differentiation: Formation of the Primitive Endoderm and Ectoderm

The second differentiation event in mammalian embryogenesis is also characterized by the appearance of an epithelial layer—the primitive endoderm—on the free surface of a group of nonpolarized cells—the ICM (see Fig. 15, below). The remaining core of ICM cells then becomes organized into a layer known as the primitive ectoderm. (The primitive endoderm is also known as the hypoblast and the primitive ectoderm as the epiblast or embryonic ectoderm. The terms primitive endoderm and primitive ectoderm will be given preference here.) The differentiation of the primitive endoderm begins around 4.0 days p.c., shortly before implantation, when there are only 20–40 cells in the ICM. Because of the small number of cells involved, it has so far been very difficult to make precise statements about the sequence of cellular and molecular changes involved in this differentiation and how they are related to the cell cycle, and to intercellular communication and organization. (See Gardner 1983 and Hogan et al. 1983 for recent reviews of primitive endoderm and ectoderm differentiation.)

There is good evidence from injection chimera experiments, using glucose phosphate isomerase (GPI) as a lineage marker (see Table 9, below), that primitive endoderm cells do not colonize the endodermal tissues of the fetus, but only the extraembryonic parietal and visceral endoderm in the yolk sacs surrounding the developing embryo (Gardner 1982, 1983). Similar experiments have shown that the primitive ectoderm lineage gives rise to the ectodermal, mesodermal, and endodermal tissues of the fetus, to the germ cells, and to the mesodermal components of the extraembryonic membranes and placenta (Gardner and Rossant 1979). These lineages are summarized in Figure 2.

Lineage Markers Used with Mouse Embryos

Until recently the only genetically determined lineage markers available for use with mouse embryos have been the GPI allozymes GPI[a] and GPI[b]. These differ in electrophoretic mobility and can only be assayed in tissue homogenates (see, e.g., Gardner and Rossant 1979). Their sensitivity and precision have therefore been limited. However, a variety of markers are now available that can be used at the cellular level on tissue sections. These are given in Table 9.

The Primitive Ectoderm Lineage

The inner core of apparently unpolarized primitive ectoderm cells gradually becomes organized into a simple epithelium surrounding a central cavity (Fig. 15). The formation of this proamniotic cavity probably involves the death of some central cells, since it frequently contains cell debris. The primitive ectoderm cells are joined by apical desmosomal junctional complexes, and have a polarized, subapical concentration of cytokeratin polypeptides (Jackson et al. 1981). They are attached via their basal surface to a thin basal lamina laid down between the ectoderm and visceral endoderm (Wartiovaara et al. 1979; Leivo et al. 1980).

Table 9 Cell Autonomous Markers Used to Follow Cell Lineages during Mouse Development

Marker	Method used	Reference
1. Monoclonal antibodies specific for *H-2^b* and *H-2^k*	histochemical staining with alkaline phosphatase and peroxidase	Ponder et al. (1983)
2. Satellite DNA sequence distribution between *M. musculus* and *M. caroli*	in situ hybridization	Siracusa et al. (1983); Rossant et al. (1986)
3. Null mutation in cytoplasmic malic enzyme (*Mod-1^n* vs. *Mod-1$^+$*)	histochemical staining	Gardner (1984)
4. Carbohydrate polymorphism recognized by *Dolichos biflorus* agglutinin	histochemical staining	Schmidt et al. (1985)
5. Injection of horseradish peroxidase	histochemical staining	Cruz and Pedersen (1985)
6. Monoclonal antibody (OX7) specific for the Thy-1.1 allele of Thy-1, a surface glycoprotein on T lymphocytes and certain other tissues, including fibroblasts and embryonic brain cells; has been used to follow grafts of embryonic into adult brain; congenic pairs of Thy-1.1 and Thy-1.2 mice are available	histochemical staining	John et al. (1972); Morris and Barber (1983)
7. High and low activity alleles of β-glucuronidase; use limited to central nervous system	histochemical staining	Mullen (1977)

Primitive Ectoderm Cells Divide Rapidly

A careful study has been made by Snow (1977, 1978a) of the growth rate of the primitive ectoderm (epiblast) in Q strain (random bred) mice. As shown in Table 11, this tissue contains about 120 cells at 5.5 days p.c. and 660 cells 1 day later, when gastrulation begins. The mesodermal cells that exfoliate through the primitive streak during gastrulation are derived from the primitive ectoderm. In order to account for the increase in cell numbers, first in the primitive ectoderm by itself, and then in both the ectoderm and mesoderm, the mean cell cycle time of the primitive ectoderm cells must be around 4.4 hours at between 6.5 and 7 days p.c. (Table 12). In fact, Snow presents evidence that the division rate is not uniform throughout the primitive ectoderm but is significantly faster (estimated cycle time

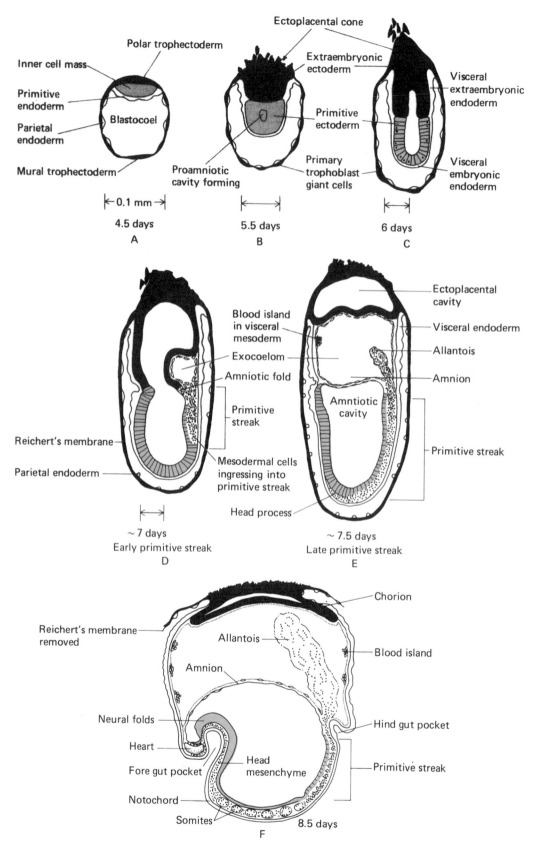

Figure 15 Schematic representation of early postimplantation development of the mouse embryo (timing based on C3H/He inbred line).

Table 11 Total Cell Numbers in the Embryonic Germ Layers

Age (days p.c.)	Number of embryos (no. of litters)	Endoderm[a]	Mesoderm[a]	Epiblast/ectoderm[a]
5.5	14(5)	95	—	120
6	6(3)	130	—	250
6.5	13(5)	250	—	660
7	7(2)	430	1220	3290
7.5	16(5)	680	6230	8060

Table reprinted, with permission, from Snow (1977).
[a]Number of cells.

Table 12 Mean Cell Cycle Times Required to Account for the Growth of the Epiblast

Age (days p.c.)	5.5		6		6.5		7		7.5
Number of cells	120		250		660		4510		14,290
Number of divisions		1.04		1.32		2.71		1.58	
Mean cell cycle		11.5		9.1		4.4		6.7	

Table reprinted, with permission, from Snow (1977).

Table 13 Estimated Cell Cycle Times (hours) for Various Regions of the Embryo

Age (days)	6.5	7	7.5
Epiblast/ectoderm	4.8	7.2	8.1
Mesoderm		22.2	13.9
Proliferative zone	2.2	3.2	3.6
Other epiblast	5.1	7.5	8.5

Table reprinted, with permission, from Snow (1977).

2–3 hours) in a group of ectoderm cells which he calls the proliferative zone (PZ) (Table 13). The mesoderm, on the other hand, appears to divide relatively slowly, with a cycle time of 20 hours (Table 13).

At Least Some Cells in the Primitive Ectoderm Are Pluripotent and Can Give Rise to Teratocarcinomas

Results from a number of different experiments suggest that at least some of the cells of the primitive ectoderm before about 7.0 days p.c. are unrestricted in their developmental potential and can give rise to a variety of different tissues if they are transferred to a new environment. For example:

1. Gardner and Rossant have isolated single primitive ectodermal cells from 4.5-day blastocysts and injected them into host blastocysts of a different GPI genotype. The primitive ectoderm cells give rise to descendants in many different embryonic and extraembryonic tissues (Gardner and Rossant 1979).

2. Blastocysts cultured in vitro can give rise to cell lines that can differentiate into many different tissues, either as teratocarcinomas or as chimeras with normal embryos (for review, see Evans and Kaufman 1983; for isolation of such lines see Section F). One possibility is that all cells in the ectoderm of the late blastocyst are equally able to give rise to pluripotent stem cell lines. An alternative is that this transformation is restricted to those primitive ectoderm cells destined to give rise to primordial germ cells. At present this question remains unresolved.

3. When preimplantation embryos of suitable mouse strains are explanted to ectopic sites, they develop more or less normally to the early egg cylinder stage (equivalent to about 6 days p.c. of normal development) and then become disorganized and generate either a teratoma or a teratocarcinoma containing undifferentiated stem cells. Similarly, LT strain parthenogenones in the ovary also develop to the 6-day egg cylinder or even early primitive-streak stage before becoming disorganized as teratomas or teratocarcinomas (for review, see Stevens and Pierce 1975; Martin 1980). Primitive ectoderms isolated from 6-day mouse embryos and transplanted to ectopic sites also give rise to teratomas or teratocarcinomas containing a wide variety of differentiated tissues from all three germ layers, as well as to undifferentiated carcinoma stem cells known as embryonal carcinoma (EC) cells (Diwan and Stevens 1976). Again, it is not known whether all cells of the primitive ectoderm can give rise to EC stem cells or only those that differentiate into primordial germ cells.

In contrast to the experiments described above, there is evidence that by 7.5 days p.c. the embryo is no longer able to give rise to teratocarcinomas containing EC cells in ectopic sites (Stevens 1970) and the cells are restricted in their developmental potential. Evidence for this comes from transplantation studies, and from the results of culturing fragments of the 7-day p.c. egg cylinder in vitro (Snow 1981).

Size Regulation

If double-sized embryos are made by aggregating morulae, (Section C), the offspring are normal size. The downshift in cell numbers takes place between about 5.5 and 6.5 days p.c. and occurs in both trophectoderm and ICM derivatives at about the same time (Lewis and Rossant 1982).

X Chromosome Inactivation

X chromosome inactivation first occurs in the trophectoderm and extraembryonic endoderm. In both these tissues it is the paternal X that is inactivated preferentially (Takagi and Sasaki 1975; Kratzer et al. 1983; Lyon and Rastan 1984). One way of looking at this is to postulate that passage through the male germ line imprints the paternal X chromosome with increased sensitivity to the inactivation process. Alternatively, the maternal X chromosome may be imprinted with relative resistance. These alternatives are discussed by Lyon and Rastan (1984). It is only in the primitive ectoderm that random inactivation of paternal and maternal X chromosomes takes place. Using two different techniques it has been concluded that this random X inactivation is probably complete by 5.5 days p.c. (Rastan 1982; Mc-Mahon et al. 1983).

Gastrulation and the Formation of Mesoderm Cells

In the mouse embryo the process of gastrulation begins at about 6.5 days p.c. when there are about 1000 cells in the primitive ectoderm (Table 11). Cells delaminate from the epithelial layer of the primitive ectoderm and accumulate as a layer of individual mesoderm cells between the primitive ectoderm and visceral endoderm. Very little is known about the mechanism controlling the detachment of mesoderm cells, but it presumably involves changes in cell–cell adhesion molecules, cell polarity, and cytoskeletal organization. Franke et al. (1982) have shown that early mesoderm cells express vimentin intermediate filaments but not cytokeratins or desmosomes. In contrast, the primitive ectoderm cells do not express vimentin, but have cytokeratins associated with desmosomal tight junctions (Jackson et al. 1981).

Morphogenetic Movements Associated with Gastrulation and the Generation of a Segmented Pattern in the Mesoderm

The process of gastrulation results in the gradual establishment of a sequence of about 65 paired blocks of mesodermal cells, the somites, along the anterior–posterior body axis. For a discussion of when this axis is established, see Smith (1985). These somite blocks eventually differentiate into vertebrae, ribs, muscles, and dermis of the skin. Since the somites are laid down in a simple linear pattern along the body axis, and give rise to major components of the body plan, the mouse is considered to be a segmented animal (Hogan et al. 1985). One approach to understanding the basic principles of gastrulation and somitogenesis in the mouse is to draw analogies with the chick embryo. If one imagines that the cup-shaped primitive ectoderm and visceral endoderm of the mouse embryo are folded out flat, they resemble the disc-shaped blastoderm of the chick embryo, which has an upper epithelial layer of epiblast cells overlying a sheet of hypoblast (Figs. 16 and 17).

The formation of the mesoderm in the chick and the appearance of a segmented pattern within it have been studied for many years (see, for example, Balinsky 1975). The most recent and detailed analysis using stereo electron microscopic techniques has been the work of Meier and his colleagues (Meier 1979, 1981). Tam has made similar studies on mouse embryos and all the evidence points to the sequence of events being essentially the same as in the chick (Tam 1981; Meier and Tam 1982; Tam and Meier 1982).

In the chick, delamination of mesoderm cells starts at the posterior end of the epiblast and then extends anteriorly to form a primitive streak or groove. Gastrulation is, in fact, not just a local event but involves extensive movement within the whole epiblast sheet, so that cells are swept into the primitive streak from the outer regions of the embryonic disc, to emerge as lateral and paraxial mesoderm cells (Figs. 16 and 17).

The next landmark of gastrulation is the appearance of Hensen's node at the anterior end of the primitive streak. Those epiblast cells that are swept through Hensen's node are laid down along the midline as a strip of mesoderm cells which

CHICK

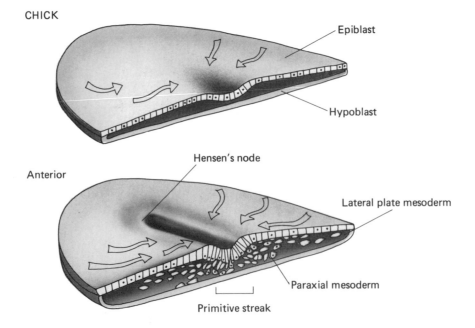

Epiblast

Hypoblast

Hensen's node

Anterior

Lateral plate mesoderm

Paraxial mesoderm

Primitive streak

MOUSE

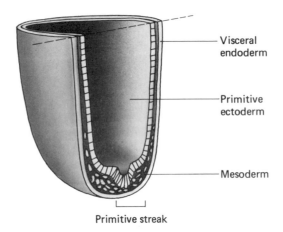

Visceral endoderm

Primitive ectoderm

Mesoderm

Primitive streak

Figure 16 Schematic representation of gastrulation in the chick and mouse embryo. (*A*) Chick blastoderm before and after formation of the primitive streak and Hensen's node. Arrows mark the direction of migration of cells within the epithelial sheet of the epiblast. Mesoderm cells are delaminating from the epiblast and accumulating between the upper and lower epithelial sheets. (*B*) Primitive streak-stage mouse embryo in section. If flattened out it would resemble the chick blastoderm above.

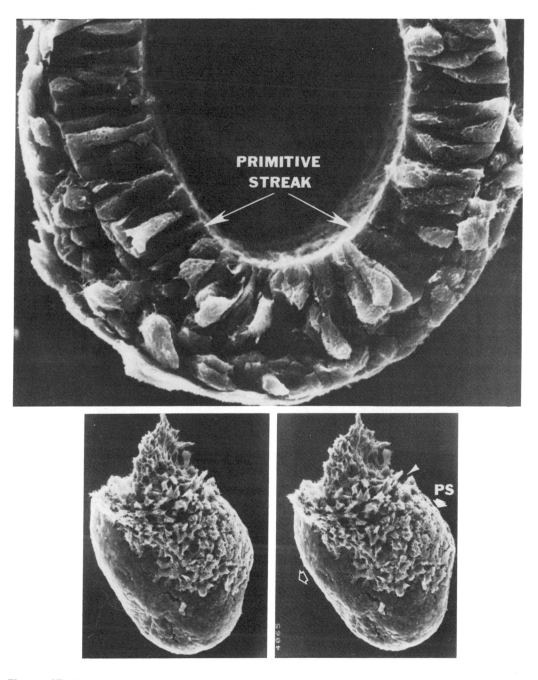

Figure 17 (*Top*) Transverse cross-fracture through the primitive streak of an early primitive streak-stage embryo showing the ingression of epiblast cells and the lateral spread of mesoderm cells from the midline. (*Bottom*) Stereo pair showing the cellular organization of the newly formed mesoderm. Arrow marks anterior. (PS) Primitive streak. (Photograph provided by Dr. Patrick Tam, The Chinese University of Hong Kong.)

differentiate into the notochord. The position of the node is not static but gradually moves in a posterior direction; as it regresses it leaves behind a trail of notochord cells separating the mesoderm on either side. Two important developments follow. One is the formation of the neural folds in the overlying ectoderm in response to inductive signals from the underlying mesoderm (neural induction). The other is condensation of the mesoderm cells on either side into paired cylindrical blocks, or somites. These changes do not take place instantaneously, but progress over several hours (in the chick) or days (in the mouse) during which extensive cell proliferation takes place (Tables 11–13). Thus, at 8.5 days p.c. new mesoderm cells are still being generated by ingression through the posterior end of the primitive streak (or later in the tail bud region) at a time when the first somites are condensing behind the regressing node. There is also an anterior–posterior gradient in the maturation of the somites. The region between the most mature, condensed somites and Hensen's node is called the segmental plate (in the chick) or the presomitic mesoderm (in the mouse). Although under the light microscope this region appears not to be segmented, analysis of carefully fixed embryos by stereo scanning electron microscopy has shown that in both organisms the mesoderm is arranged into paired whorls of cells, or somitomeres (Figs. 18 and 19) (Meier 1979; Tam and Meier 1982). In the mouse there are about six somitomeres in the presomitic mesoderm. If this region is dissected away and cultured in vitro these presumptive somites condense and mature in the absence of signals from the rest of the embryo (Tam et al. 1982). Somitomeres have been seen in the very earliest primitive streak-stage embryos (Tam et al. 1982). It is assumed that little or no cell mixing takes place between the mesodermal whorls, but this question cannot be resolved until a good cell marker is available for this stage of development.

In the mouse the first somite blocks are visible under the light microscope at about 7.75 days p.c. (Theiler 1972) when there are about 25,000 cells in the ectoderm plus mesoderm (Snow 1977). The first five somites remain small and disperse to give rise to a few of the muscles of the eye and neck (Balinsky 1975; Noden 1983). (Most of the mesoderm of the head and face is derived from the cephalic neural crest.) As will be described below, the remaining somites differentiate into a variety of tissues, including vertebrae and muscles. This differentiation gradually obscures the segmented pattern of the embryo, and in the adult the only trace is in the sequence of vertebrae and ribs down the spinal column. Each vertebra has a unique morphology, and classically the series has been divided into five regions: cervical (7), thoracic (13), lumbar (6), sacral (4), and caudal (30–31). The final number shows very little variability within an inbred strain, except in the caudal region. The number of cells initially allocated to each somite is not identical, and starts at about 150–200 in the cervical region, increases to 1000–1500 in the lumbar and sacral regions, and then declines progressively in the tail somites (Tam 1981).

The final maturation and differentiation of the somites in the body region involves a complex series of changes in cell adhesion and movement. The cells first become epithelial, with their inner apical surfaces enclosing a central cavity and their basal surfaces resting on an outer basal lamina of extracellular glycoproteins, including type I collagen and laminin (Leivo et al. 1980). Each somite block then becomes subdivided into three regions: the sclerotome, the dermatome, and the myotome (Fig. 20). The fate of cells in each region is:

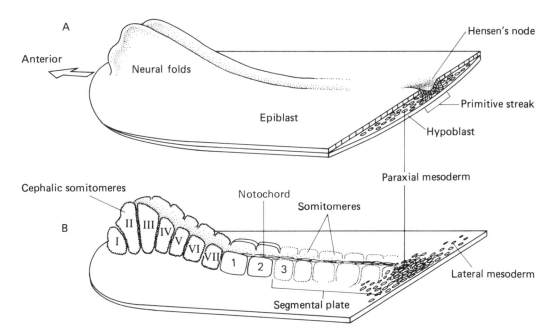

Figure 18 Schematic representation of a neural fold-stage chick or mouse embryo. (*A*) Viewed with ectoderm in position. (*B*) The ectoderm and neural tube have been removed to reveal the underlying mesoderm. (1–3) The first somite blocks to condense; (I-VIII) the cranial somitomeres detected only by stereo scanning electron microscopy. In vivo there are up to six somitomeres in the segmental plate of the mouse embryo.

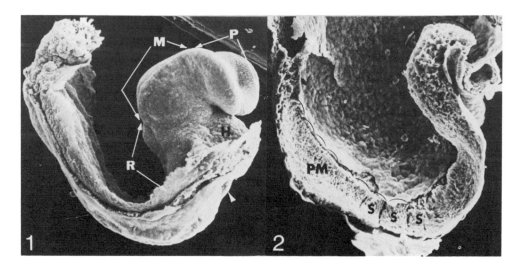

Figure 19 Scanning electron micrograph of the neural fold-stage mouse embryo (∼8.5 days p.c.). (*1*) External morphology showing headfold with prosencephalon (P), mesencephalon (M), and rhombencephalon (R) regions of the prospective brain. White arrowhead is foregut and H is heart. (*2*) Left half of an embryo from which the neuroepithelium has been removed. There are three somites (S) and seven somitomeres in the cranial region and six in the presomite mesoderm (curved arrows).

Somite block region	Tissue
Sclerotome	vertebrae (the anterior sclerotomes of one pair of somites and the posterior sclerotomes of the preceding pair cooperate to form one vertebra)
	ribs in the thoracic regions
Dermatome	connective tissue in the dermis of the skin
	dermal component of the hair follicles
Myotome	muscles in the body wall; the myotomes of the four to five somites adjacent to the fore and hind limb buds give rise to muscles in the limbs

Apart from the somite-derived tissue, the kidney of the mouse also has a segmental history, since it develops in the abdominal region from serially repeated blocks of mesodermal cells (the nephrotomes) connected to the somites. In the mouse, the lateral plate mesoderm gives rise to a variety of tissues, including the mesenchyme of the limb bud (and eventually limb cartilage and bone). In mammals, the lateral plate mesoderm is not obviously divided into segmental blocks (for discussion, see Hogan et al. 1985).

Origin of the Definitive Endoderm

There is good evidence that the endoderm of the fetus (the so-called definitive endoderm) in contrast to the extraembryonic endoderm of the yolk sacs is derived

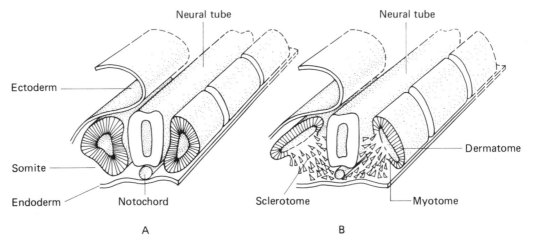

Figure 20 Schematic representation of the differentiation of the somites in the trunk region of the mouse embryo. The ectoderm has been pulled back on one side to reveal the somites adjacent to the neural tube. (*A*) The somite blocks have an epithelial organization. (*B*) Some cells migrate around the notochord to form the sclerotome cells of the vertebrae. The remainder differentiate into dermatome and myotome.

from the primitive ectoderm. This evidence comes from blastocyst injection experiments (e.g., Gardner and Rossant 1979), ectopic transplantation of primitive ectoderm cells (Diwan and Stevens 1976), and experiments designed to follow the fate of [³H]thymidine-labeled cells grafted into primitive streak-stage embryos in culture (Beddington 1981). However, very little precise information is available about the morphogenetic movements involved in the formation of the endoderm in vivo, presumably from cells moving through the anterior end of the primitive streak (the head process) (Green 1975; Tam and Meier 1982). Later in development, tissues such as the liver and lungs are derived from the definitive endoderm (see Fig. 2).

Toward a Fate Map of the Mouse Embryo

Gastrulation in the mouse embryo involves extensive morphogenetic movements during which cells in the epithelial primitive ectoderm layer are swept toward and through the primitive streak, to emerge as mesodermal and endodermal cells. In the chick, interpretation of the complex morphogenetic movements of gastrulation has been aided by mapping experiments on the living embryo in which grafts of cells marked with [³H]thymidine (Nicolet 1970) or with vital dyes have been used to trace the fate of different regions of the epiblast. The resulting fate map is very crude, but delimits patches of epiblast cells which will, in the normal embryo, end up as different tissues of the body, e.g., somites, gut, notochord, neural tube, etc. (Balinsky 1975; Slack 1983). These maps say nothing about the time at which the epiblast cells and their descendants become irreversibly committed to their developmental fate and acquire their specific positional addresses or epigenetic codings (for discussion, see Slack 1983).

The small size of the mouse embryo, its more complex culture requirements, the absence of a good histochemical cell marker, and the rapid dilution of [³H]thymidine label due to cell division conspire to make fate mapping and cell lineage studies a formidable task. Nevertheless, some experiments have been done using primitive ectoderm cells labeled with [³H]thymidine (Beddington 1981). In addition, Snow (1981) and Tam and Snow (1980) have studied the regionalization of the 7- and 7.5-day p.c. embryo by excising pieces of the egg cylinder, culturing them in vitro, and scoring them for the different tissues they generate. They find that even at 7 days p.c. there are regions within the ectoderm and mesoderm that are restricted in their developmental repertoire.

Further references to the problem of timing of cell commitment in the primitive streak-stage mouse embryo and the generation of a segmented pattern in the mesoderm can be found in Hogan et al. (1985). This review also describes some mouse mutants that have defects in the organization of somite-derived repetitive structures, and discusses the significance of the recent observation that a short, highly conserved protein-coding DNA sequence, the homeo box, located in a set of *Drosophila* genes involved in segment number and identity is also present in the mouse and human genomes (McGinnis et al. 1984).

Development of the Nervous System

The brain and spinal cord (central nervous system) develop from the neural tube. This, in its turn, develops from the neural plate, a thickening of the ectoderm overlaying the notochord and paraxial mesoderm, which gradually folds up to form a hollow tube (Fig. 21). The closure of the neural tube starts at about the seven-somite stage (8.25 days p.c.), at the level of the fourth and fifth somite. It then proceeds anteriorly and posteriorly in a zipper-like manner. Closure of the anterior neuropore is complete by the 15- to 20-somite stage (9 days p.c.), while closure of the posterior neuropore is not complete until the 32-somite stage (10 days p.c.) (Theiler 1972). The initial formation of the neural plate is induced by interaction of the ectoderm and the underyling mesoderm. This process of neural induction has been studied for many years in several species of vertebrates (for references, see Slack 1983), but the molecular basis is not yet known. Many forces are involved in the folding of the simple epithelial layer of the neural plate, including cell elongation (mediated by microtubules) and apical constriction (mediated by microfilaments) (for review, see Karfunkel 1974). The early neural tube also shows a series of swellings known as neuromeres which in the body region are intersomitic (Morriss-Kay 1981; Tuckett et al. 1985) (Fig. 22). The precise forces generating neuromeres and whether the constrictions separating them represent barriers to cell migration are problems under investigation.

The peripheral nervous system develops from the neural crest. These cells detach from the crests of the neural folds as the lateral walls fuse. The individual cells accumulate in the midline and then migrate in streams along the matrix-rich tracts between the somites. From there they diverge along a number of pathways

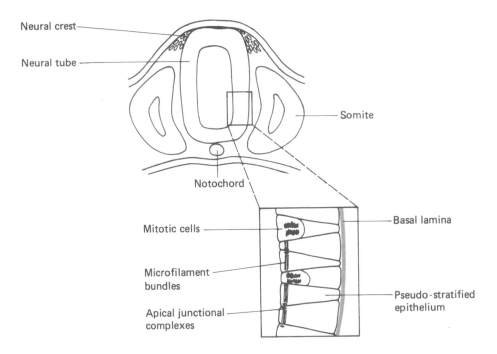

Figure 21   Schematic section through an 8.5-day p.c. mouse embryo at about the level of somite 4 showing the neural tube and the early neural crest. The inset shows the organization of the neural tube before extensive cell proliferation has generated multiple layers of cells. Here they form a simple, pseudostratified epithelium surrounded by a basal lamina. Mitotic figures are always found adjacent to the lumen of the tube.

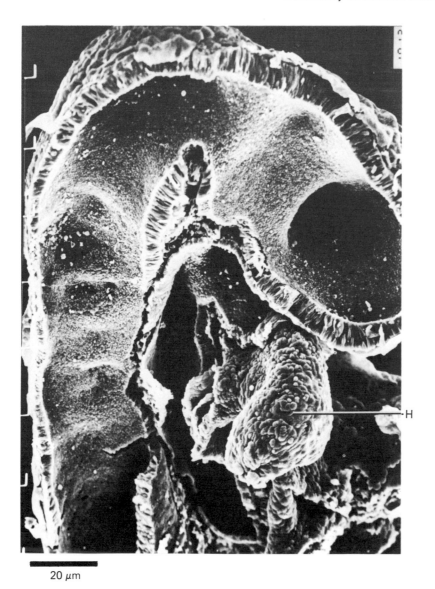

20 μm

Figure 22 Scanning electron micrograph of a sagittally-halved, 12-somite rat embryo showing the apical surface of the neural epithelium. Only the cranial region is shown, to illustrate the pattern of neural segments or neuromeres. (H) Heart. (Photograph provided by Dr. Gillian Morriss-Kay, University of Oxford.)

and eventually differentiate into a variety of tissues (e.g., peripheral ganglia, facial muscle and cartilage, pigment cells), depending on the environment in which they come to lie (Le Douarin 1982).

Extraembryonic Tissues

The extraembryonic tissues form an integral part of the life support system essential for the maintenance, nourishment, and protection of the fetus within the uterus. They constitute the placenta, the parietal yolk sac (parietal endoderm and trophoblast), the visceral yolk sac (visceral endoderm and mesoderm), and the amnion (mesoderm and ectoderm) (Fig. 23). Studies on gene expression in extraembryonic tissues have focused on proteins that are synthesized at high levels in order to fulfill certain specialized functions. Examples cited below are the serum component α-fetoprotein (AFP), which is secreted by the visceral endoderm, and the extracellular matrix glycoprotein laminin, which is laid down by the parietal endoderm. Other studies have revealed more esoteric properties of these tissues. For example, in the trophoblast and the parietal and visceral endoderm (but not in the visceral mesoderm or amnion), the paternal X chromosome is specifically inactivated (Takagi and Sasaki 1975; Kratzer et al. 1983; Lyon and Rastan 1984), and both repetitive and single-copy DNA sequences are undermethylated (Chapman et al. 1984). In addition, in all extraembryonic tissues, and in particular the amnion, there is a high level of expression of the proto-oncogene, c-*fos* (Muller et al. 1983; Curran et al. 1984). The physiological significance of these various observations is not yet known.

Extraembryonic Endoderm: The Primitive Endoderm Gives Rise Only to Visceral and Parietal Endoderm

The endoderm cells of the yolk sacs surrounding the mouse embryo are derived from a precursor pool of about 20 bipotential primitive endoderm (PrE) cells (Fig. 15). Lineage studies with GPI allozyme-marked cells have shown that the primitive

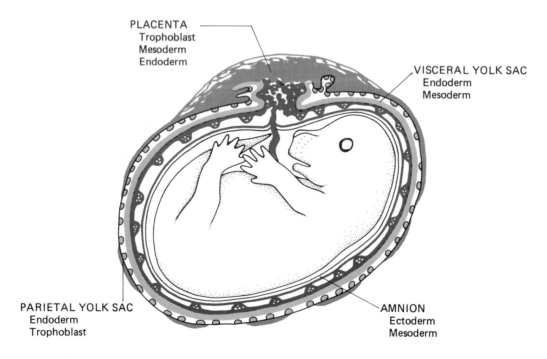

Figure 23 The placenta and extraembryonic membranes of the 13.5-day embryo.

endoderm does not contribute to the definitive endoderm of the adult mouse (Gardner and Rossant 1979; Gardner 1982, 1983), but only to visceral and parietal extraembryonic endoderm. This lineage study has been confirmed using the Mod-1^+/Mod-1^n marker system (Gardner 1984).

PrE cells first differentiate on the blastocoelic surface of the ICM at 4–4.5 days p.c. and can be distinguished from the 20 or so primitive ectoderm cells by a number of morphological features (Nadijcka and Hillman 1974), in particular a more extensive endoplasmic reticulum which is swollen with secretory material (presumably including type IV collagen, laminin, and fibronectin; see later). The PrE cells do not, however, form a well-defined, polarized epithelium at this stage (see, for example, Enders et al. 1978). As the primitive and extraembryonic ectoderm layers grow and elongate to form the core of the egg cylinder (see Fig. 15), the outer endoderm cells differentiate into two morphologically and biochemically distinguishable subpopulations, the visceral endoderm (VE) and the parietal endoderm (PE). Endoderm cells that remain in contact with the egg cylinder become organized into a distinct epithelium of VE cells, with apical desmosomal junctions and microvilli, and many small and larger vacuoles and organelles distributed in the cytoplasm in a polarized way. VE cells surrounding the primitive ectoderm are known as visceral embryonic endoderm, and tend to be flatter or squamous, whereas those surrounding the extraembryonic ectoderm (visceral extraembryonic endoderm) are taller and have a very vacuolated cytoplasm and numerous microvilli (Fig. 24) (Hogan and Tilly 1981). A completely different morphology is seen in the parietal PE cells (Figs. 24 and 25). These first appear at the time of implantation (Enders et al. 1978) when primitive endoderm cells migrate on to the inner surface of the trophectoderm which is covered by a thin basal lamina containing fibronectin and laminin (Wartiovaara et al. 1979; Leivo et al. 1980). In contrast to the VE cells, PE cells are individual and migratory, do not form specialized intracellular junctions, do not have an obvious polarity in the distribution of their intercellular organelles, and coexpress vimentin and cytokeratins (Lane et al. 1983; Lehtonen et al. 1983b). The most significant feature of PE cells is their enormously enlarged endoplasmic reticulum filled with secretory material, including components of the thick basement membrane (Reichert's membrane) which is laid down by the PE cells (see below). In vivo, the apical surfaces of PE and VE cells are closely apposed (Fig. 23) and the limited space between the two is filled with secretions of both the PE and VE cells and substances from the maternal circulation that have been filtered through Reichert's membrane. A discussion of the role of cell interactions in the differentiation of primitive endoderm into visceral or parietal phenotypes can be found in Hogan and Tilly (1981), Hogan et al. (1983), and Gardner (1982, 1983).

Gene Expression in Visceral Endoderm

The most important functions of the visceral endoderm are: (1) absorption, i.e., uptake of substances from the maternal circulation that have filtered through Reichert's membrane into the cavity of the parietal yolk sac and (2) secretion, i.e., production and secretion of serum proteins and other substances required by the fetus such as AFP, transferrin, high- and low-density apolipoproteins, and α_1-antitrypsin. Transcripts hybridizing to an insulin gene probe have also been detected in the rat yolk sac (Muglia and Locker 1984). The visceral endoderm therefore performs some of the same functions as both the large intestine and fetal liver (Meehan et al. 1984).

The morphology of visceral endoderm cells is highly specialized for absorption since the cells have numerous apically located microvilli and coated pits, as well as lysosomes, etc. (Figs. 24 and 26). Since the cells are polarized, it is likely that the absorptive functions are located at the apical surface, whereas secretion of AFP, transferrin, and other serum components may take place via the basal surface, but this question has not been resolved. In the visceral yolk sac, the basal surface is adjacent to the mesoderm layer containing fetal blood vessels. The di-

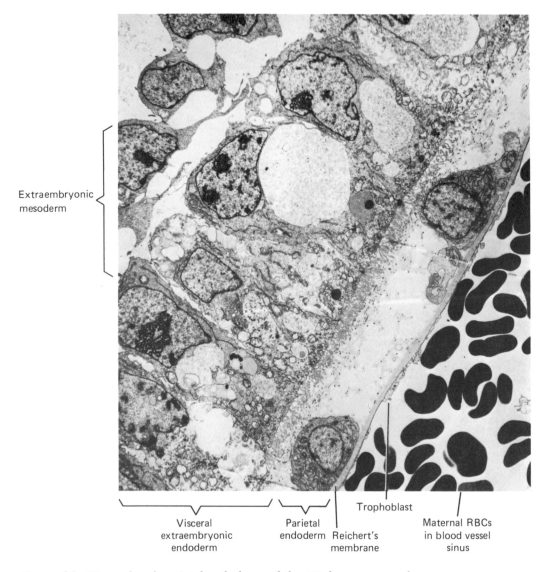

Extraembryonic mesoderm

Visceral extraembryonic endoderm   Parietal endoderm   Reichert's membrane   Trophoblast   Maternal RBCs in blood vessel sinus

Figure 24   Visceral and parietal endoderm of the 7.5-day mouse embryo.

rection of secretion of other VE products such as plasminogen activator is a matter of speculation.

Some of the products of the visceral endoderm are given in Table 10. The most specific "marker" so far recognized is AFP, a glycoprotein of M_r 68,000 made only by the visceral endoderm, or the fetal or regenerating liver. It represents approximately 25% of the total protein synthesis of the visceral endoderm in 15.5-day mouse embryos and its mRNA represents 15% of total poly(A)$^+$ RNA (Andrews et al. 1982a,b; Janzen et al. 1982). The precise function of AFP is unknown, but since it is the major γ-globulin in fetal blood it may fulfil the same role as serum albumin in adult blood. The AFP gene is closely related to the albumin gene, probably by duplication and divergence of a common ancestral sequence (Gorin and Tilghman 1980; Eiferman et al. 1981).

By immunoperoxidase staining of sections of mouse embryos of different stages, AFP is first detected in the visceral embryonic endoderm at 7 days p.c. (Dziadek and Adamson 1978; Dziadek and Andrews 1983). It is absent from visceral endoderm around the extraembryonic ectoderm as a result of the inhibitory influence of this tissue. If visceral extraembryonic endoderm is separated from the underlying extraembryonic ectoderm, the cells will start to synthesize AFP within 12 hours (Dziadek 1978). Recent work with in situ hybridization using [^{3}H]cDNA probes to AFP has confirmed the results of the earlier metabolic labeling experiments and also shows that AFP mRNA is present in all VE cells of the 14-day visceral yolk sac and not in the mesodermal cells (Dziadek and Andrews 1983).

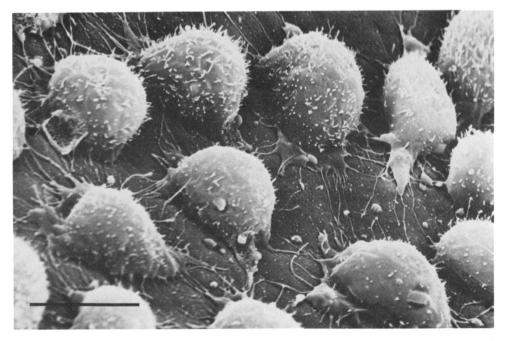

Figure 25 Scanning electron micrograph of parietal endoderm cells of 10.5-day p.c. rat embryo attached to Reichert's membrane. Scale bar, 10 μm. (Photograph provided by Dr. Stephanie Ellington, Department of Physiology, University of Cambridge.)

Table 10 Markers of the Extraembryonic Endoderm of the Normal Mouse Embryo

Marker synthesized[a] or expressed	Primitive endoderm (reference number)	Parietal endoderm of the 10- to 14-day embryo	Visceral endoderm of 6- to 8-day egg cylinder		Visceral endoderm of the 10- to 14-day yolk sac
			visceral embryonic endoderm	visceral extraembryonic endoderm	
Fibronectin	+(1)	-(2,3)	+(1)	+(1)	+(2,3)
Type-IV procollagen	+(1,4)	+ + +(5,6,7)	+(8)	+(8)	+(8) – (4,9)
Type-I collagen	?	-(8)	-(8)	-(8)	+(4,9)
Type-III collagen	-(8)	-(8)	-(8)	-(8)	-(4,9)
Laminin	+(8)	+ + +(5,6,10)	+(8)	+(8)	+(8)
Cytokeratins	? but see (11)	+(12,22,23,24)	+(13,22)	+(13,22)	+(21,23,24)
Desmosomes	-(14)	-(13,15,16)	+(13)	+(13)	+(9,16)
α-Fetoprotein	-(17)	-(17)	+(17)	-(17) +(18) modulated	+ +(17,19)
Transferrin	-(20)	-(20)	+(20)	+(20)	+(20)
Vimentin	?	+(12,24)	-(13)	-(13)	-(24)
Plasminogen activator	?	+(25) 79K tissue activator	?	?	48K urokinase (25)
α₁-Antitrypsin	?	—	?	?	+(26)
Apolipoproteins	?	—	?	?	+(26)

Citations are not meant to be exhaustive and give only the most recent publications in which reference to earlier work can be found. Where appropriate, biosynthetic labeling experiments and mRNA translation studies are given precedence over immunofluorescence localization studies.

1. Wartiovaara et al. (1979); 2. Amenta et al. (1983); 3. Hogan et al. (1980); 4. Adamson and Ayers (1979); 5. Hogan et al. (1980); 6. Smith and Strickland (1981); 7. Kurkinen et al. (1982); 8. Leivo et al. (1980); 9. Clark et al. (1982); 10. Cooper et al. (1981); 11. Kemler et al. (1981), for isolated inner cell masses cultured in vitro; 12. Lane et al. (1983); 13. Jackson et al. (1981); 14. Nadijcka and Hillman (1974); 15. Enders et al. (1978); 16. Hogan and Cooper (1982); 17. Dziadek and Adamson (1978); 18. Dziadek (1978); 19. Janzen et al. (1982); 20. Adamson (1982); 21. R. Kemler (pers. comm.), visceral yolk sac is positive for Troma 1 staining; 22. Boller and Kemler (1983); visceral endoderm around eight-day egg cylinders is Troma 1 +ve, Troma 3 –ve; parietal endoderm cells are Troma 1 +ve and Troma 3 +ve; 23. Oshima et al. (1983); positive staining with anti-EndoA and EndoB sera; 24. Lehtonen et al. (1983b); 25. Marotti et al. (1982); 26. Meehan et al. (1984).

Table is adapted from Hogan et al. (1983).

[a]When immunocytochemical studies alone are cited, synthesis is inferred from positive cytoplasmic staining. However, in many cases with antibodies to matrix glycoproteins this staining is only weak, as judged by the authors themselves.

Gene Expression in Parietal Endoderm

The most obvious feature of parietal endoderm cells is the fact that they are specialized for synthesizing and secreting a thick basement membrane known as Reichert's membrane between themselves and the trophectoderm. Until about 16 days of gestation (when it breaks down), Reichert's membrane is one of the major barriers between the maternal and fetal environments. This is because in the rodent the endothelial cells of the maternal blood vessels break down to give large sinuses and the trophectoderm cells lying between these blood sinuses and the embryo do not remain as a continuous shell below the placenta but gradually die away. Although it is assumed that Reichert's membrane acts as a passive filter, keeping out maternal cells and large molecules, there is in fact little hard information about its permeability properties and even less is known about the rela-

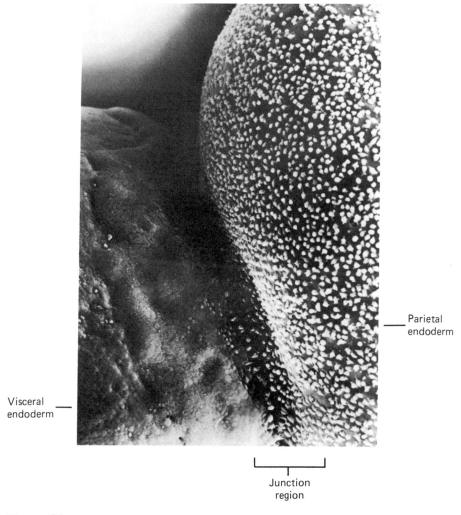

Parietal endoderm

Visceral endoderm

Junction region

Figure 26 Junction region between parietal and visceral endoderm in the 7.5-day mouse embryo. The parietal endoderm on Reichert's membrane has been folded back to reveal the underlying visceral endoderm layer around the egg cylinder. Magnification, 440×.

tionship between the function of Reichert's membrane and its structure and composition. PE cells synthesize large amounts of basement membrane components, including laminin, entactin, type IV procollagen, and heparan sulfate proteoglycan (Table 9). The mature cells do not synthesize fibronectin (Hogan 1980; Smith and Strickland 1981; Amenta et al. 1983) and it is not considered to be a structural component of Reichert's membrane (Semoff et al. 1982). However, it is likely that primitive endoderm cells do make fibronectin and possibly use it in migration on the trophectoderm (Hogan et al. 1983). RNA from PE cells has been used to isolate cDNA clones for both type IV procollagen and laminin (Kurkinen et al. 1983a,b; Barlow et al. 1984) and Reichert's membrane is proving to be an excellent model system for studying basement membrane synthesis, assembly, and remodeling (Hogan et al. 1984). PE cells also synthesize large amounts of tissue plasminogen activator, which has a different molecular weight, antigenicity, and inhibitor sensitivity to the urokinase-type plasminogen activator made by VE (Marotti et al. 1982).

Differentiation of the Extraembryonic Mesoderm

Some of the mesoderm cells generated by the early primitive streak contribute toward the following different extraembryonic tissues (see Fig. 15 and Gardner 1983).

1. Amnion. This is generated from both embryonic ectoderm and mesoderm. It appears first as a fold and then a continuous roof over the top of the cup-shaped primitive ectoderm (Fig. 15). It then expands rapidly and with the turning of the embryo forms a thin membrane surrounding the fetus. The ectodermal and mesodermal cells of the amnion have very different morphologies and are separated by a basement membrane (Fig. 27).

2. Allantois. This starts as a finger-like projection of mesodermal cells from the posterior margin of the embryonic ectoderm where the primitive streak first arises. It expands upwards, fuses with the chorion, contributes to the placenta, and gives rise to the blood vessels of the umbilical cord.

3. Mesoderm of the chorion and visceral yolk sac. Mesodermal cells generated from the posterior primitive streak migrate onto the inner surface of the visceral endoderm and give rise to the first hematopoietic tissue of the fetus in the form of blood islands in the visceral yolk sac. Mesodermal cells that cover the chorion also contribute to the placenta.

The Structure and Function of the Placenta

By midgestation the placenta has become a very complex organ consisting of both fetal and maternal tissues and blood cells. The development of the placenta is described in detail by Theiler (1972) and only a schematic representation of two stages is given in Figure 28. An important feature distinguishing the mouse placenta from the human is that the maternal blood vessels break down, so that the blood cells come into direct contact with the fetal trophoblast.

The major fetal tissues of the mouse placenta are:

1. Trophoblast. In the outer spongiotrophoblast layer closest to the uterine decidual tissue, most of the trophoblast cells are diploid, whereas in the inner labyrinth layer many of the cells are polyploid and giant. In both layers there are maternal blood sinuses containing maternal blood cells, and even after dissection from the uterus the outer layer is also contaminated with maternal decidual cells (Rossant and Croy 1985).

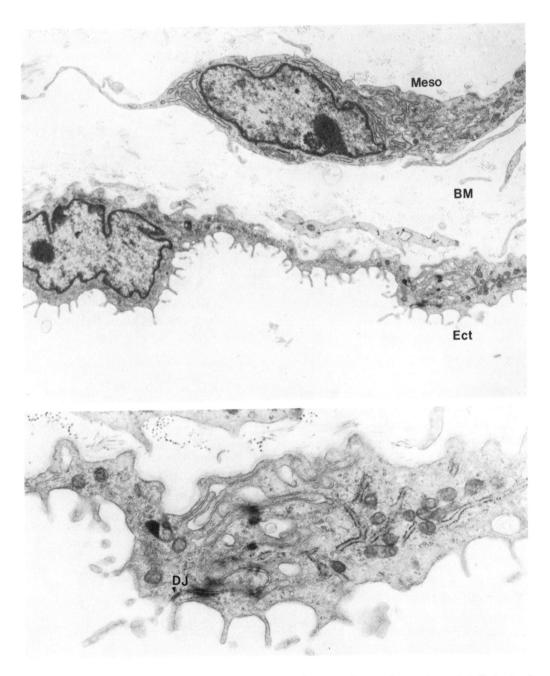

Figure 27 (*Top*) Section through the amnion of a 10.5-day embryo. (Meso) Cell derived from mesoderm; (Ect) cell derived from ectoderm; (BM) basement membrane; (DJ) desmosonal junction. The mesodermal cells are not joined by specialized junctions. (*Bottom*) Higher magnification of an ectodermal cell (Ect) in top panel.

2. Mesoderm. After the allantois fuses with the chorion it gives rise to fetal capillaries and blood vessels. These mingle with the trophoblast cells and with maternal blood sinuses in the labyrinth.

3. Visceral and parietal endoderm. The crypts of Duval contain both visceral and parietal endoderm. The visceral cells synthesize AFP and the parietal cells make basement membrane.

Apart from playing an important function in the transfer of nutrients and metabolites into and out of the fetal circulation, the placenta synthesizes many steroid, polypeptide, and prostaglandin-type hormones that are required for the coordination of maternal and embryo physiology during pregnancy (see, for example, Soares et al. 1985).

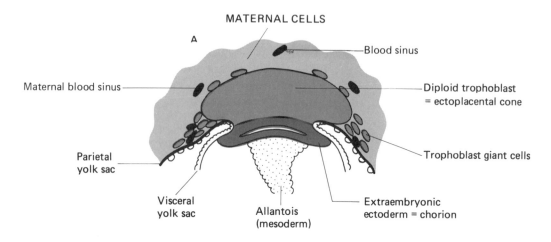

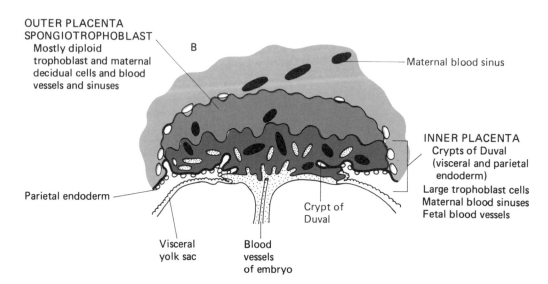

Figure 28 Schematic representation of the mouse placenta at 8.5 days p.c. (*A*) and 14.5 days p.c. (*B*). Note that in the older embryo, both the outer and inner placenta have a significant contribution of maternal cells. The best source of trophoblast tissue free of maternal contamination is, therefore, the 7.5- to 8.5-day p.c. ectoplacental cone (see Rossant and Croy 1985).

THE ADULT MOUSE

Mouse Coat Color and Its Genetics

As discussed in the History section, the genes affecting coat color were among the first to be tested for Mendelian inheritance in mice. Since these early studies, more than 50 genes have been identified that affect hair growth and pigmentation. Here, only the briefest outline of the subject will be given to stimulate the curiosity of people handling mice for the first time. For a full and scholarly account of the genetics of coat color, readers are referred to books by Silvers (1979) and Grüneberg (1952).

During embryogenesis, each hair follicle develops from an invagination of the epidermis, the base of which surrounds a condensation of mesodermal cells known as the dermal papilla (Fig. 29). In the mouse there are about 50 hair follicles/mm²,

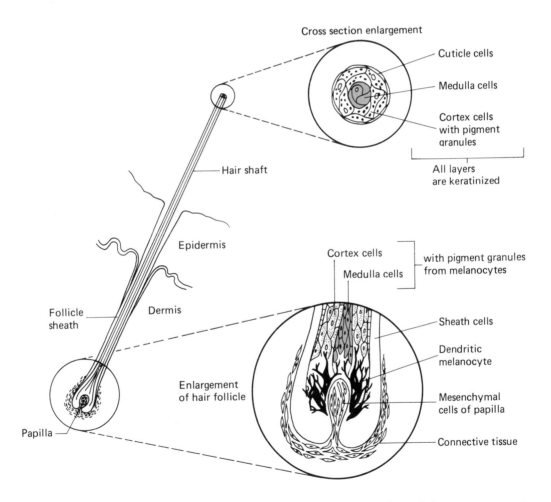

Figure 29 Schematic representation of a hair follicle to show the relative arrangement of cells derived from the epidermis (sheath, cortex, medulla), the dermis (mesenchymal cells of papilla and connective tissue), and the neural crest (melanocyte).

or about 500,000 over the entire skin (Potten 1985), and they are formed between 14 days p.c. and 3 days after birth. Melanocyte precursor cells (melanoblasts) migrate from the neural crest into the hair follicles and take up a position above the dermal papilla (Fig. 29). There are about 20–30 dendritic melanocytes in each follicle and evidence suggests that they are derived by division from one founder melanoblast (Potten 1985). Each melanocyte has the capacity to synthesize two kinds of pigment, pheomelanin (yellow) and eumelanin (black or brown). Both pigments are derived from tyrosine and involve an initial conversion with the copper-containing enzyme tyrosinase. After this initial step, different enzyme systems generate the alternate chromatophores, which are then linked to proteins and incorporated into pigment granules of different size and shape. During active phases of the hair growth cycle the melanocytes secrete these granules, which are taken up by the cortical and medullary epidermal cells of the hair shaft. Several different kinds of hair are found on the body, besides the whiskers (vibrissae); the three kinds of larger overhairs are the monotrich, awl, and auchene, while the more abundant, smaller underhairs are known as zigzags. Their morphology is determined by the mesodermal component of the follicle.

Since the genes affecting hair color were among the first to be studied, the loci were named alphabetically. Here only the first four will be discussed; they are A (agouti), b (brown), c (albino), and d (dilute). It is generally considered that the wild-type alleles in the European house mouse are A^+, B^+, C^+, D^+ (Grüneberg 1952).

A (agouti)

This is a complex locus (on chromosome 2) with 17 alleles reported. Hair from an agouti mouse is black with a subapical band of yellow. This pattern is generated by the mesodermal component of the hair follicle which transiently inhibits the production of black pigment by the melanocytes during an early phase of the growth cycle. Nonagouti (a/a) mice are therefore plain black (e.g., C57BL) or brown (C57Br), apart from a few yellow hairs on the ears and around the genitals. The effect of agouti varies over different parts of the body, so that on the underside and ears the yellow banding is more extensive.

The mutation yellow (A^y) has been particularly well studied since it is a homozygous lethal, with A^y/A^y embryos dying around implantation. The heterozygote A/A^y is yellow, but black pigmentation of growing hairs can be induced by injection of α-melanocyte-stimulating hormone (α-MSH) (see Silvers 1979 for references). The A^y mutation is closely associated with an ecotropic murine leukemia virus (MLV) genome, suggesting that it is caused by retrovirus insertion (Copeland et al. 1983).

b (brown)

The wild-type allele at this locus, B, produces black eumelanin, whereas the most recessive allele, b, produces brown. A/A, b/b mice are known as cinnamon, a color produced by their yellow-banded brown hairs. The *b* locus is thought to control the size and shape of the pigment granule as well as the chemical structure of the eumelanin pigment deposited within it.

c (albino)

The wild-type (*C*) allele is dominant over all mutations at this locus. These mutants result in a deficiency or alteration in the structure of tyrosinase, and do not affect the number or distribution of the melanocytes. Albino mice (*c/c*) have no pigment at all either in the coat or eyes (Fig. 30) while other mutants at the *c* locus have altered pigmentation, e.g., chinchilla *c^h* and chinchilla mottled *c^m* (See Fig. 30). It is important to realize that not all "albino" mice have the same coat color genes. In the presence of *c/c* the effects of changes at the *A*, *b*, and *d* loci are masked. In albino mice, as in all albino mammals, there is also a defect in the visual pathways, with a greater degree of crossing over of the optic nerves in the brain.

d (dilute)

The dilute locus is one of the class of genes affecting coat color through an alteration in the morphology of the pigment granules in the melanocytes. In *d/d* mice the pigment granules are clumped and the melanocytes are less dendritic than those in wild-type *D/D* mice. The molecular basis for this effect is not known, but there is good evidence that the *d* mutation in DBA/2J mice is caused by integration of an ecotropic MLV retrovirus genome (Jenkins et al. 1981; Rinchik et al. 1985). Another mutation that causes abnormal coat color through an effect on melanosome morphology is *bg* (beige). The *bg/bg* mouse is a model of the Chediak-Higashi syndrome in humans and is associated with a defective morphology of membrane-bound intracellular organelles such as lysosomes and melanosomes (see Fig. 30).

Spotting Mutations

A particularly interesting class of pigmentation mutants are those known as "spotting" (e.g., *W^v* [see The Origin of Germ Cells and Their Migration to the Genital Ridges]), piebald (*s/s*), and belted (*bt/bt*). Examples of the first and last are shown in Figure 30. Spotting genes affect the migration, viability, or differentiation of melanoblasts, the precursors of the melanocytes that arise from neural crest cells (see Development of the Nervous System). Between 8.5 and 9 days of gestation neural crest cells are still located near the neural tube; by 11 days p.c. they have reached the skin of the trunk and by 12 days, the limbs. Genes affecting melanocyte migration are therefore more likely to result in lack of pigment in the hair follicles on the ventral surface and extremities (e.g., piebald), since the cells have to migrate further to reach these locations.

Figure 30 (*A*) CBA×C57BL/10 F$_1$ hybrid. This mouse is heterozygous agouti/nonagouti (*A/a*) and homozygous black (*B/B*) and therefore shows the dominant agouti phenotype of black hairs with a subapical yellow band (see Section A, Mouse Coat Color and Its Genetics). The F$_2$ will segregate both black, nonagouti (*a/a*), and agouti (*A/a, A/A*) offspring in the ratio 1:3. The ears have been clipped as part of a numbering system (see Section B). An agouti, black mouse is generally described as "brownish-grey" and this is the basic color of wild *M. musculus domesticus* mice.

(*B*) Nonagouti C57BL/6 (*a/a*; +/+) (*left*) and a coisogenic mouse homozygous for the beige mutation which arose in the Jackson Laboratory (*a/a*; *bg^J/bg^J*) (*right*). Coisogenic strains differ only at one locus, as a result of a mutation arising in an inbred strain. Beige is a recessive mutation that causes clumping and fusion of intracellular melanosomes in both the melanocytes of the skin and the choroid of the eye (which are derived from the neural crest), and the retinal epithelium (which is derived from the neurectoderm). The overall result is lightening of coat and eye color. Another effect of beige is the fusion of lysozomes in a wide variety of cell types, including liver, visceral yolk sac endoderm, proximal kidney tubules, and leukocytes. One consequence is that *bg/bg* mice have defective natural killer cells and abnormal secretion of lysosomal enzymes. The mutation has been widely used as a model for the Chediak-Higashi syndrome in humans (Silvers 1979; Brandt et al. 1981).

(*C*) Homozygous nude mouse (*nu/nu*) (outbred background). Nude is a recessive mutation affecting the ectodermal component of the thymus, producing a small, cystic gland which does not provide an environment for T-cell maturation. The skin has the normal number of hair follicles but the hairs are abnormally keratinized and break off at the skin surface. Homozygous *nu/nu* mice are widely used in immunological research and for the establishment of transplantable human tumors (Nomura et al. 1977).

(*D*) Albino (*c/c*). The albino locus probably encompasses the structural gene for tyrosinase. The recessive mutation *c* at this locus blocks tyrosinase activity, possibly by producing inactive enzyme. Note that the mutation is expressed in both melanocytes and retina. Albino mice were recorded in Greek and Roman times and the *c* locus was the first to be studied for Mendelian inheritance in mice (see Historical Perspective section).

(*E*) Chinchilla mottled, an example of an unstable gene and position-effect variegation. Chinchilla (*c^{ch}*) is a recessive mutation at the albino locus (*c*) which reduces tyrosinase activity and the deposition of yellow and, to a lesser extent, black pigment. The chinchilla mottled mutation (*c^m*) arose in the progeny of a neutron-irradiated wild-type male. The mouse shown here is homozygous (*c^m/c^m*) and has a fine-grained variegated pattern, the darker patches corresponding to melanocyte clones phenotypically chinchilla (*c^{ch}*) and the lighter patches to clones phenotypically extreme chinchilla (*c^e*) or albino (*c*). The switch must occur late in development because the patches are small and finely intermingled. Lightly pigmented cells also occur in the retina, but only comprise 1% of the total, hence the dark eyes. There may be an environmental or positional effect on the expression of *c^m*, as the belly hairs of *c^m/c^m* are invariably lighter than dorsal hairs, and the unpigmented cells of the retina are clustered around the optic nerve. One suggestion is that the *c^m* mutation resulted in the gene being shifted near to heterochromatin, which in some cells extends into the gene and partially inactivates it (Deol and Truslove 1981). Other explanations are possible, however.

(*F*) Dominant spotting. A large series of mutations has been recorded at the dominant white spotting locus (*W*). All of them affect the proliferation and migration of three different cell populations; the neural crest-derived melanocytes, the hematopoietic stem cells, and the primordial germ cells. Mutations can therefore cause defects in epidermal but not retinal pigmentation, as well as anemia and infertility. The mutations are expressed in the migratory cells themselves, rather than in the environment into which they move, and the phenotype can vary depending on genetic background. The mouse

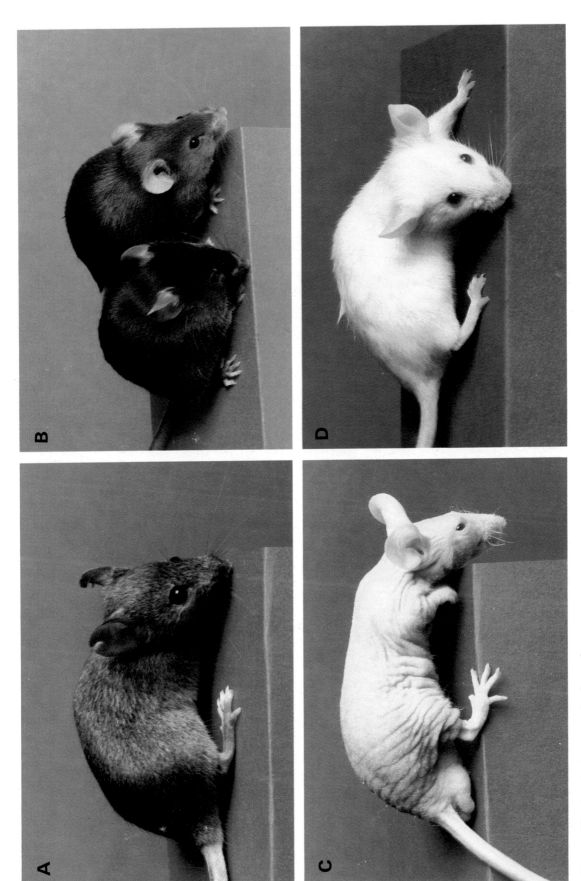

Figure 30 (See facing page for legend.)

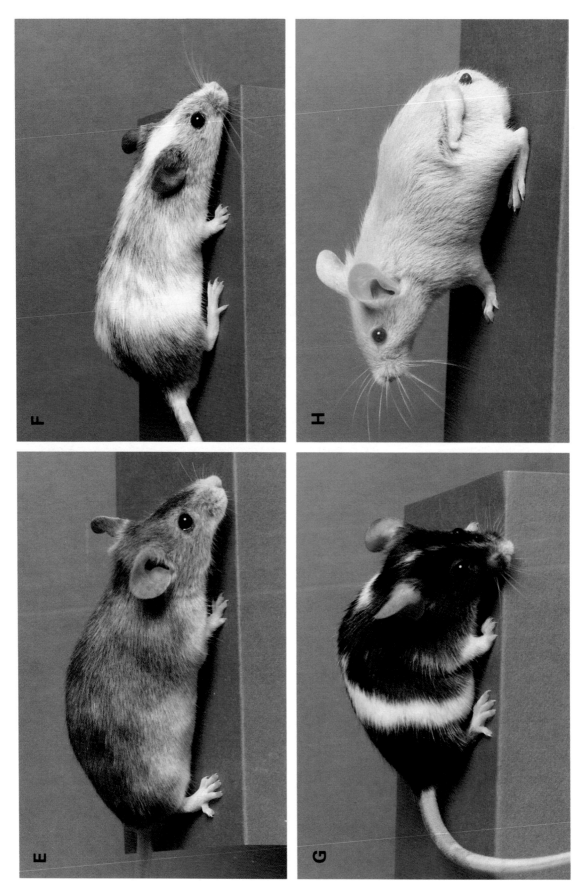

Figure 30 (continued) (*See facing page for legend.*)

illustrated here is heterozygous for the allele extreme spotting ($W^e/+$) and is on an agouti (A/A)[C3H/HeH × 101/H]F$_1$ background (Loutit and Cattanach 1983). Note the extensive nonpigmented (and therefore melanocyte-free) patches on the flank and back, and the white blaze on the forehead. Homozygous W^e/W^e mice die soon after birth from anemia. They are completely white with black eyes.

(G) Belted (*bt*) is another spotting mutation. The white patches are areas which the melanocytes fail to reach by migration, or in which they cannot survive or proliferate (Silvers 1979). The homozygote (*bt/bt*) here is on a nonagouti (*a/a*) background that has been selected for increased expression of the belted phenotype, so that the mouse has white spots on the head and shoulders in addition to the classical "belt."

(H) This mouse is homozygous for pudgy, pink-eyed dilution, chinchilla, and agouti (*pu/pu*; *p/p*; c^{ch}/c^{ch}; *A/A*). Homozygous pudgy mice have extensive defects in the axial skeleton and ribs. In particular, there is a reduction in the number of caudal vertebrae, which are also very disorganized. The limbs, however, are normal. Pudgy embryos can be recognized as early as 9 days p.c. In the caudal region the paraxial mesoderm condenses, but does not become divided up into discrete somite blocks (Grüneberg 1961). Pink-eye dilution (*p*) is, like albino, one of the old mutations of the mouse fancy. Seventeen alleles are known at the *p* locus (Silvers 1979). *p* alters melanosome structure and drastically reduces the deposition of black and brown pigment but only has a small effect on the synthesis of yellow pheomelanin in both melanocytes and retinal epithelium. Melanosome structure is altered. Mottled (p^m) and unstable (p^{un}) alleles exist in which the effect of the *p* mutation is variegated. In p^m the effect of the gene is modified, while in p^{un} there appears to be a high rate of somatic reversion to wild type, the rate varying with age and genotype (Silvers 1979; Green 1981).

(*A*, *B*, *C*, *D*, and *H* are from the National Institute for Medical Research, Mill Hill, London. *B* and *H* originated from the Jackson Laboratory. *E*, *F*, and *G* are from the MRC Mammalian Development Unit, London. *F* was provided by Dr. Mia Buehr and *E* by Dr. Ian Jackson, but both mutations originated in the MRC Radiobiology Unit at Harwell, U.K.) (Photographs by Mike Tatham, National Institute for Medical Research, London.)

Section B

SETTING UP A COLONY FOR THE PRODUCTION OF TRANSGENIC MICE

To generate transgenic mice successfully and efficiently, the investigator must routinely be able to obtain both a large number of embryos for microinjection and an ample supply of pseudopregnant recipients. In addition, the health of the animals must be maintained and good breeding practices must be followed. This section describes a mouse colony suitable for transgenic experiments that is based on colonies at the Memorial Sloan-Kettering Cancer Center (E.L.), Columbia University (F.C.), and the National Institute for Medical Research (B.L.M.H.). The strains, ages, and numbers of mice we describe will, of course, need to be modified according to each investigator's experimental requirements. Although this section is aimed primarily at researchers who want to establish a transgenic mouse system, the information it contains will also be useful for those who want to perform other types of studies utilizing mouse embryos. In this section, we do not cover breeding strategies that are required for more genetically sophisticated experiments, such as gene mapping or the derivation of congenic strains, etc. For this information, the reader is referred to Green (1975) and Foster et al. (1983). In addition, we do not give a detailed description of the reproductive physiology of mice; this can be found in Whittingham and Wood (1983).

This section was written with the assumption that the investigator has access to an animal facility that has provisions for cage washing, food and water supply, is equipped with proper ventilation, temperature, humidity and light control, and has access to veterinary care.

The animals maintained in a colony for the production and analysis of transgenic mice can be divided into five categories, each of which will be discussed here separately:

1. Female mice for matings to produce fertilized eggs.

2. Fertile stud male mice.

3. Sterile stud male mice for the production of pseudopregnant females.

4. Female mice to serve as pseudopregnant recipients and foster mothers.

5. Transgenic mice, including "founder" mice and transgenic lines derived from these founders.

Female Mice for Matings to Produce Fertilized Eggs

The first decision to be made when setting up a colony for the production of transgenic mice is the strain that will be used as donor of the eggs for microinjection. For certain experiments, such as the transfer of one allele of a mouse gene into a strain carrying a different allele (Grosschedl et al. 1984), genes must be introduced into a defined genetic background; in such studies zygotes from inbred strains of mice are generally used for microinjection. To date the inbred strain most widely used has been C57BL/6J. If the genetic background is not critical to the experiment, injections are most often performed with F_2 hybrid zygotes generated from matings between F_1 hybrid male and female mice [e.g., (C57BL/6J×CBA/J)F_1 female×(C57BL/6J×CBA/J)F_1 male]. F_2 hybrid zygotes derived from several different F_1 hybrids have been used successfully to produce transgenic mice. These F_1 hybrids include: C57BL/6J×SJL, C57BL/6J×CBA/J, C3H/HeJ×C57BL/6J, C3H/HeJ×DBA/2J, and C57BL/6J×DBA/2J. In general, the generation of transgenic mice and their subsequent breeding is more efficient when F_2 zygotes are used for microinjection (Brinster et al. 1985), because of the relatively poor reproductive performance of inbred mice.

Superovulated females (see Section C, Superovulation) are used almost exclusively over naturally ovulating females (see below) for the production of fertilized eggs. On average, 20–30 eggs can be recovered from a 4- to 6-week-old superovulated C57BL/6J (abbreviated B6) or B6×CBA F_1 female, whereas only 6–8 and 8–10 eggs, respectively, can be recovered from naturally ovulating B6 and B6×CBA F_1 female mice. Even higher numbers of eggs can be obtained by superovulation if younger females are used, but these must be specially reared, as described in

Section C (Superovulation). Superovulation minimizes the labor and expense involved in obtaining sufficient numbers of eggs for microinjection. For example, to recover approximately 200 fertilized eggs, one would need to dissect oviducts from 20–30 naturally ovulating females but only from 7–10 superovulated females. Consequently, the number of female mice that must be maintained for the isolation of eggs is minimized by the use of superovulation.

Table 14 indicates the number of female mice used and the number of eggs recovered in a typical microinjection experiment using superovulated C57BL/6J female mice. Ten 4- to 6-week B6 females are injected with pregnant mare's serum (PMS) at 1–2 pm (assuming a light period of 5 am–7 pm) and with human chorionic gonadotropin (hCG) 46–47 hours later. After the administration of hCG each female mouse is placed in a cage with a stud B6 male (one female and one male per cage). Typically, 6–10 of the females will mate, to yield a total of 120–300 viable eggs. Usually between 75 and 90% of these will be fertilized, as evidenced by the presence of two pronuclei. Two polar bodies are also usually visible (see Section A). The remaining 10–25% will have no pronuclei; these represent eggs that have not been fertilized (one polar body) or that have already undergone pronuclear breakdown in preparation for the first cleavage (two polar bodies). Approximately 60–80% of the zygotes that are microinjected will survive; these are reimplanted into the oviducts of pseudopregnant foster mothers, 30–35 embryos per mouse. In a typical day, two to five oviduct transfers are performed. The figures listed in Table 15 (Section C) for superovulated C57BL/6J eggs are very similar to those that would be obtained with B6×CBA F_2 eggs; the main advantage of F_2 zygotes is the fraction of transferred embryos that develop to term: 7–10% for B6 eggs and 15–25% for F_2 eggs (see Table 15).

If the strain to be used as an egg donor is commercially available, then it is generally more convenient (and often less expensive) to purchase females of breeding age rather than to raise them in the colony. Many inbred strains and several F_1 hybrids are available from commercial suppliers, and one can arrange to receive a weekly shipment of mice at a specified age. If microinjection experiments are to be performed only sporadically during the year, then it definitely makes more sense to buy females when they are needed rather than to maintain a breeding colony continually. On the other hand, if the desired strain or hybrid is not commercially available, then it is necessary to produce the mice in one's own facility. In addition, young females at the optimal age and weight for maximal egg yield by superovulation can rarely be obtained commercially and must be specially bred.

To produce a supply of female mice, for example, B6×CBA F_1 hybrids, 40–50 breeding cages should be maintained, each containing one male (CBA) and one

Table 14 A Typical Microinjection Experiment with C57BL/6J Superovulated Eggs

Number of females set up to mate	Number of plugs obtained	Total number of eggs isolated	Total number of eggs with injectable pronuclei	Number of eggs surviving injection	Number of oviduct transfers
10	6–10	120–300	90–240	60–160	2–5

female (B6). When the F_1 progeny reach 3 weeks of age, they are weaned and separated by sex. For maximal production, one litter should be weaned before the next litter is born, or the older pups may trample the younger ones and prevent them from feeding. Such a breeding colony should yield 20–30 female progeny per week (and an equivalent number of males, most of which would be discarded). It is possible to increase the yield of female progeny weaned per mating cage by placing two females with each male and culling the male progeny at birth. Note that it takes 3–4 weeks from the time a breeding colony is set up to the time the first progeny are born, and another 4–8 weeks before they reach reproductive age. Therefore, such a breeding colony must be maintained even when the progeny are not immediately required, because it takes several months to establish a new colony.

It is important to wean the progeny before they become sexually mature or else the female progeny will mate with the father. Males that are weaned together before maturity (from the same or separate cages) may be kept together indefinitely, as long as they are not exposed to females. Mature males that have been kept in separate cages, or have been exposed to females, must never be placed in the same cage or they will fight. Females generally do not fight, and may be caged with other females at any time.

A few simple procedures will allow one to maintain a breeding colony at peak efficiency. First, the breeding cages should be checked a few times a week and the date of birth and size of each new litter recorded on the cage card. The presence of this information on the cage card will immediately indicate when a litter should be weaned. In addition, if the size of a litter at weaning is significantly smaller than at birth, or if the sizes of the litters produced by a mother decrease (i.e., from 8–10 to 2–4) with her increasing age, then the breeding pair should be replaced. Second, the male should be left in the cage with the female. The reason for this is that immediately after giving birth a female mouse goes into postpartum estrus and will mate with the male. Thus, a good breeding pair will produce a litter every 3 weeks. Third, the breeding pairs should be replaced about every 6 months or even earlier if the female has not produced a litter for 2 months. Fourth, when a litter is weaned and separated by sex, the date of birth should be recorded on the new cage card; also mice of significantly different ages (greater than a week) should not be mixed. This is important as only females in a certain age range can be used for efficient superovulation. In addition, to keep a colony healthy, it is best to keep younger mice in preference to older ones; this can only be done if the date of birth is known.

Fertile Stud Male Mice

A transgenic mouse facility that uses 150–200 embryos 5 days a week will require about 50 fertile stud males. Each stud male must be housed in a separate cage to avoid fighting and injury. Males should be placed in separate cages 1–2 weeks before being presented with a superovulated female, because the dominant male in a litter may suppress testosterone synthesis, and consequently sperm production, in his littermates. Male mice reach sexual maturity between 6 and 8 weeks of age and can be used as studs for 1–2 years depending on the strain. In practice, however, inbred males should be replaced at 8–10 months of age, as their reproductive performance tends to decrease after this time.

Each superovulated female is placed *individually* with a stud male, and is checked the next morning for a copulation plug (see Section C, Natural Matings, for a description of the copulation plug). The mating or "plugging" performance of the male is monitored by recording on each male's card the date a female was presented and the presence or absence (+ or –) of a plug the next morning. A normal male will plug a superovulated female nearly every time one is presented; if a male fails to plug a superovulated female several times in a row, or if his plugging average is less than 60–80%, he should be replaced.

It has been observed that the sperm count will be depressed for several days after a male has mated. Therefore, to insure that the maximum number of eggs is fertilized, a stud male should not be used for about a week after plugging a female. This requires that more stud males be maintained than would be used on any one night. For example, if one wants to obtain 150–200 eggs on a given day, about 10 superovulated females would be mated with 10 stud males; if this number of eggs is required 5 days a week, it is necessary to maintain 50 stud males, so that each set of 10 males is only used once per week.

Sterile Stud Male Mice for the Production of Pseudopregnant Females

Sterile males are required for matings to generate pseudopregnant recipients and are usually produced by vasectomy (see Section C). Alternatively, genetically sterile males may be used, such as mice that are doubly heterozygous for two recessive lethal haplotypes of the *t* complex. For vasectomy, males of at least 2 months of age from any strain with a good breeding performance (such as an outbred or F_1 strain) will be suitable. Before using a vasectomized male in an experiment, it should be tested for sterility by mating it with females to obtain at least three plugs. If the operation was successful, none of the plugged females should become pregnant.

To produce four to eight pseudopregnant females 5 days a week, 20 sterile males should be sufficient. Vasectomized males may be set up in matings every night, but preferably only every other night. Our experience indicates that they tend to maintain a high plugging rate for up to 2 years. A plugging record (as described above for the fertile males) should be kept for each sterile male. Any mouse that fails to plug a female four to six times in a row should be replaced.

Female Mice to Serve as Pseudopregnant Recipients and Foster Mothers

Pseudopregnant mice are prepared by mating females in natural estrus (see Section C, Vasectomy and Generation of Pseudopregnant Females) with vasectomized or genetically sterile males. The females should be at least 6 weeks of age and, ideally, weigh ≥ 20 g; both outbred mice and F_1 hybrid mice make suitable recipients. The very best recipient would be a female who has already successfully reared a litter (an "experienced mother"). In our experience, B6×CBA F_1 females work well as pseudopregnant recipients and subsequently make excellent mothers, rearing litters as small as two pups.

Many investigators use coat color differences between the strain of the egg donor and the strain of the pseudopregnant recipient to be certain that any mice born actually derive from the donor eggs. For example, microinjected C57BL/6J eggs are transferred to B6×CBA F_1 females that have been plugged by vasectomized B6×CBA F_1 males. Mice that develop from the donor eggs will all have a black coat color, whereas a F_1×F_1 mating will produce agouti and black progeny in a 3:1 ratio. Therefore the presence of any agouti progeny would indicate that the male was not properly vasectomized. However, if the vasectomized males are properly tested for sterility, the use of coat color markers is not necessary.

If two to five oviduct transfers are planned, then approximately 20 females selected to be in estrus should be placed with 10 vasectomized males, two females per male. Since not all females judged to be in estrus will actually be in estrus, placing two females with each male increases the probability that each male will produce a plug. If the numer of vasectomized males is a limiting factor (as each male must be separately caged, and maintenance charges are often assessed on a "per cage" basis), this will result in the greatest number of plugs per male. On the other hand, if the number of available females in estrus is limited (for example, if only 10 females in estrus are available on a particular day), each female should be placed with a separate male, to maximize the number of plugs produced. With properly selected females, 5–10 pseudopregnant plugs should be produced by either of the above strategies. It is always advisable to plan the matings so that an excess of pseudopregnant plugs is obtained. Often an attempt to transfer embryos (Section C, Embryo Transfers) to a particular pseudopregnant female will fail, and it will then be necessary to use a second pseudopregnant recipient.

Extra pseudopregnant females should be saved, as they can be used again in matings once they begin to cycle (10–14 days after plugging). Females that fail to mate should be placed back in a stock cage of females, and can be used again whenever they enter estrus.

In general, females enter estrus and ovulate every 4–5 days. Thus, in a colony of randomly cycling females, 20–25% of the mice should be in estrus at any time. Consequently, to obtain 10–15 females in estrus an average of 50–75 mice, and some times a greater number, must be examined. Therefore, if 50 pseudopregnant females are to be obtained per week, it is necessary to maintain a steady-state supply of at least 100 females between the ages of 2 and 5 months for pseudopregnant recipients. The rate at which new females will have to be produced (or purchased) will depend on the number of pseudopregnant females actually used per week, since unused pseudopregnant females can be recycled. As noted above, a breeding colony of 40 cages, each containing one male and one female, will typically yield 20–30 female progeny (at weaning) per week. If the females to be used as recipients are purchased from a commercial vendor, they should be given 1 week to adjust to the light/dark cycle of the colony before being used.

Sometimes only one or two of the injected eggs will develop to term in a pseudopregnant recipient; in these cases the fetuses tend to be quite large and the mother may have difficulty giving birth. To save the litter, it may be necessary to sacrifice the pseudopregnant recipient, deliver the pups by caesarian section, and cross-foster them to a new mother. (Caesarian section and fostering are described in Section C.) To be sure that a suitable mother will be available for cross-fostering, it is advisable to mate several females with fertile males so that they will give birth at the right time for use as foster mothers. In most cases, this requires setting up matings with these females 1 day later than the matings to produce pseudopregnant recipients. Most mice will give birth on day 19 of pregnancy (counting the day of the plug as day 0), although this may vary between strains. If a pseudopregnant recipient has not given birth by midday on day 20 (or 1 day later than the normal time for that strain), the litter should be delivered immediately by caesarian section. If the mother to be used for cross-fostering has been mated 1 day later than the pseudopregnant recipient, she should have given birth by that time. If the animal colony is large, so that litters are born nearly every day, it is not necessary to set up females specifically to be used for cross-fostering; instead, any female in the colony that has given birth that day can be used as a foster mother.

When cross-fostering, it is helpful to use coat color differences to distinguish the potential transgenic pups from those belonging to the foster mother. If B6 eggs are being used for microinjection, for example, a suitable mating to produce foster mothers is CBA/J male × (B6 × CBA)F$_1$ female, which will produce all agouti pups. If this is not possible, pups can be marked by clipping the end of the tail, or by clipping the toes.

Occasionally as many as 10–17 babies may be born to one mother if, for some reason, large numbers of eggs were transferred. Care should be taken to avoid some babies becoming runted as these are more likely to be sterile as adults. It may be advisable to cross-foster a few babies after a day or so to a lactating female, or to add a lactating "aunty" to the cage.

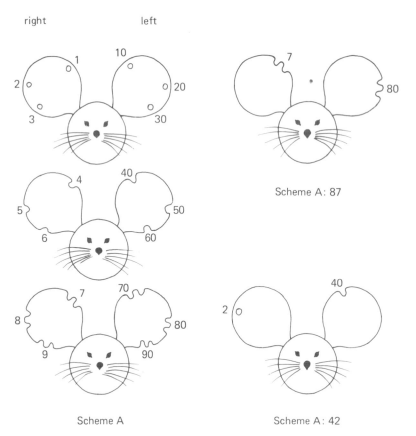

Figure 31 Two schemes for marking a mouse using continous number series.

Transgenic Lines of Mice

Potential transgenic mice are usually screened for the presence of the injected gene by performing a Southern or dot blot hybridization to DNA extracted from the tail (see Section D, Analysis of Transgenic Mice, and Section F). At the time the tail biopsy is performed on an animal, it is assigned an identification number and marked with that number. Two schemes for marking a mouse using a continuous number series are diagramed in Figure 31. Scheme A involves only ear punching and uses numbers from 1 to 99 only. Scheme B involves both toe clipping and ear punching and includes all numbers up to 9999. Such marking schemes will allow one to distinguish unambiguously each member of a transgenic colony.

Mice that develop from injected eggs are often termed "founder" mice. As soon as it has been determined that a given founder mouse is transgenic, it is usually mated to begin establishing a transgenic line. If the founder is a female, and if one wants to develop a transgenic line from her, it is necessary to wait until she has given birth and raised at least one litter before she can be sacrificed. If the founder is a male, he can be placed with two females which are checked each day and replaced with a new female as soon as each is "plugged." In this way, the male can sire many litters within a few weeks. As soon as a male has plugged a sufficient number of females (typically six to eight), he may be sacrificed, if necessary, for the analysis of gene expression. However, if one wants to be sure of establishing a transgenic line, the founder should not be sacrificed until positive transgenic progeny have been identified. Although most transgenic founders will transmit the foreign gene to 50% of their offspring, approximately 20% of transgenic founders are mosaic and transmit the gene at a lower frequency (e.g., 5–10% instead of 50%). In addition, a proportion of females that are plugged will fail to get pregnant, for example, if the male is sterile or semisterile, or will not raise a litter successfully.

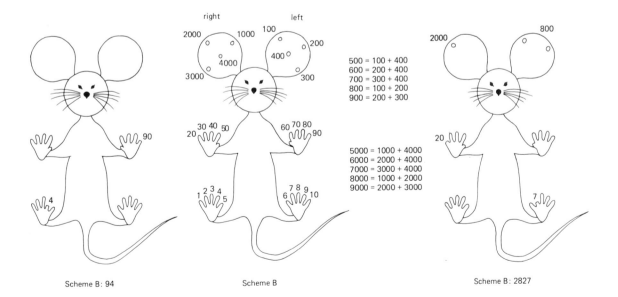

Scheme B: 94   Scheme B   Scheme B: 2827

If the founder mouse is derived from an inbred strain, it is normally mated with mice of the same inbred strain to maintain the gene on a defined genetic background. If the mouse is an F_2 hybrid, or if it is not important to maintain the gene on an inbred background, it is often more efficient to mate it with F_1 hybrid mice, which are healthier and make better mothers than inbred mice.

Homozygous transgenic mice are produced by setting up heterozygous intercrosses. One-quarter of the progeny from such a cross should be homozygous with respect to the integrated foreign DNA, one-half hemizygous, and one-quarter nontransgenic. Assays to distinguish homozygous from hemizygous animals are described in Section D. Approximately 5–15% of the DNA integration events in transgenic mice produce recessive lethal mutations; consequently not all transgenic lines can be maintained in the homozygous state.

Nomenclature for Transgenic Mice

The recommended nomenclature for transgenic mice is outlined in *Mouse News Letter* 72 (Peters 1985). Briefly, the symbol Tg should be used to indicate transgenic, followed in brackets by the chromosome into which the gene is inserted. If the chromosome is not known, then the number zero should be used. Also enclosed in the same brackets should be details of the inserted gene, e.g., Hba for an inserted α-globin sequence. If desired, the species of origin of the gene may be denoted by the chromosomes, e.g., HSA, human; MMU, mouse, etc. After the brackets, a serial number for the laboratory of investigator should be added.

Long-term Storage of Transgenic Mice by Freezing Embryos

Details for freezing preimplantation embryos and for long-term storage can be found in Whittingham (1980a) and in Leibo and Mazur (1978).

Section C

RECOVERY, CULTURE, AND TRANSFER OF EMBRYOS

This section has two functions. First, it provides information to people who want to isolate and culture mouse embryos for a variety of biochemical studies other than the introduction of new genetic information. It can therefore be used independently of Sections B and D, although readers will find the information on setting up a mouse colony and breeding mice in Section B very useful. The second function of this section is to describe the techniques of transferring embryos into the oviduct or uterus of pseudopregnant recipients. These techniques are required for a variety of experiments, including the production of transgenic mice.*

*Certain experimental procedures in this manual may be the subject of national or local legislation or agency restrictions. Users of this manual are responsible for obtaining the relevant permissions, certificates, or licenses in these cases. Neither the authors of this manual nor Cold Spring Harbor Laboratory assume any responsibility for failure of a user to do so.

To obtain fertilized eggs for microinjection or other experiments, one option is to set up natural matings, i.e., matings in which the timings of ovulation and fertilization are controlled by environmental conditions rather than by administration of gonadotropins. Females that are maintained on a constant light-dark cycle tend to ovulate once every 4–5 days, 3–5 hours after the onset of the dark period. Males that are maintained under the same conditions will copulate with females in estrus (i.e., ovulating females) at about the midpoint of the dark period. This means that fertilization takes place 1–2 hours after ovulation. For microinjection experiments a convenient lighting cycle is 7 pm–5 am dark, 5 am–7 pm light (midpoint when fertilization takes place is 12 midnight).

It is possible to identify females in estrus by examining the color, moistness, and degree of swelling of the vagina. The following table, from Champlin et al. (1973), describes the appearance of the vagina at different phases of the estrous cycle.

Stage of estrous cycle	Appearance of the vagina
Diestrus	vagina has a small opening; tissues are blue and very moist
Proestrus	vagina is gaping; tissues are reddish-pink and moist; numerous longitudinal folds or striations are visible on both the dorsal and ventral lips
Estrus	vaginal signs are similar to proestrus, but the tissues are lighter pink and less moist, and the striations are more pronounced
Metestrus 1	vaginal tissues are pale and dry; dorsal lip is not as edematous as in estrus
Metestrus 2	similar to metestrus 1, but the lip is less edematous and has receded; whitish cellular debris may line the inner walls or partially fill the vagina

To set up matings, females (6 weeks to 4 months of age) are examined in the afternoon and those in estrus are placed with males (one or two females in each cage with one male). Males reach maturity at around 6–8 weeks, depending on the strain, and once used for mating should be kept in individual cages, or they will fight. The morning after mating the females are checked for the presence of a copulation plug in the vagina (vaginal plug). This consists of coagulated proteins from the male seminal fluid and in most strains can easily be seen. In some, however, it is small and lies deep in the vagina, and can only be seen by using a probe. The vaginal plug usually dissolves about 12–14 hours after mating. Usually, 50% of the selected females will mate and each will contain between 7 and 13 fertilized eggs, depending on the strain. (Additional information on the mating behavior of mice can be found in Whitten and Champlin 1978.)

SUPEROVULATION

For experiments requiring the recovery of large numbers of preimplantation embryos, such as the microinjection of eggs, the viral infection of morula, and metabolic labeling of embryos with radioisotopes, gonadotropins are often administered to females prior to mating to increase the number of eggs that are ovulated, i.e., to induce superovulation. Pregnant mare's serum (PMS) is used to mimic follicle-stimulating hormone (FSH) and human chorionic gonadotropin (hCG) to mimic luteinizing hormone (LH). Efficient induction of superovulation in mice depends on several variables: the age and weight of the females; the dose of the gonadotropins and their time of administration; and the strain of mice. In addition, the number of superovulated eggs that actually become fertilized depends on the reproductive performance of the stud males. These important features of the superovulation procedure are discussed below.

The Influence of Age and Weight

The sexual maturity of the female is a major factor affecting the number of eggs that are superovulated. The best age for superovulation varies from strain to strain, but usually falls between 3 and 5 weeks during the prepubescent stage of development. For example, the optimal age for C57BL/6J females is 25 days (M.P. Rosenberg, pers. comm.) and for BALB/cGa mice is 21 days (Gates 1971). By this stage of development, a wave of follicle maturation has taken place that increases the number of follicles capable of responding to follicle-stimulating hormone (FSH) to a maximum. FSH is the active ingredient in PMS, which is the first hormone administered in the superovulation procedure.

Age, however, is not always a reliable indicator of the sexual maturity of a female; the nutritional status and health of a female mouse can also affect follicular maturation. Underweight and/or sick animals tend to be retarded in development and to yield a reduced number of eggs after superovulation. For example, in the C57BL/6J strain, the maximum number of viable fertilized eggs is obtained only from superovulated females that are 25 days old and weigh between 12.5 and 14 g.

Mice at this precise age and weight will usually not be available from commercial suppliers. Commercial animals are weaned in a batch at the end or beginning of a week so that their actual age will vary within a consignment. Also, the breeding conditions for commercial mice will not be designed to produce animals at a maximum weight for a given age. Thus, to obtain females at the optimal age and weight for superovulation, it is necessary to establish one's own breeding colony. Procedures for setting up a breeding colony are described in Section B. In addition to those procedures, it is also advisable to maintain the breeding pairs on a high-fat diet (Purina Mouse Chow 5020) and to cull the litters of males within 7 days of birth to ensure an ample supply of milk to the females. Such steps should enable the female offspring to obtain an optimal weight by 3 weeks.

Although female mice obtained from a commercial supplier will usually not yield the maximum number of superovulated eggs obtainable in a given strain, such mice, superovulated between 3 and 6 weeks of age, will produce a significantly larger number of fertilized eggs than will naturally ovulating females at any age (see Section B). In addition, the number of eggs produced from superovulated 3- to 6-week-old commercially supplied females will be suitable for many types of experiments, including the microinjection of eggs.

The Dose of Gonadotropins

The recommended dose of PMS for most strains is 5 IU injected intraperitoneally. (Subcutaneous injections are also equally effective.) This hormone is generally supplied as a lyophilized powder (e.g., the product sold by Organon under the code name Gestyl and by Intervet Laboratories Ltd. as Folligon). For administration, the PMS is resuspended at 50 IU/ml in sterile 0.9% NaCl and then divided into convenient aliquots. It can be stored in this form at $-20°C$ for at least a month but does not last indefinitely. A dose of 0.1 ml is injected into each animal.

hCG is the second gonadotropin that is administered to induce superovulation. It serves as a substitute for luteinizing hormone (LH) and is required for rupture of the matured follicles. Generally, injections of 5 IU are administered, although a dose of 2.5 IU may be sufficient to assure ovulation in most strains. As it is important that the hCG get into circulation quickly, before release of endogenous LH, it should be injected intraperitoneally. hCG is also supplied commercially as a lyophilized powder (e.g., from Sigma and Intervet Laboratories Ltd.). It is resuspended at 500 IU/ml in sterile water, divided into 100-μl aliquots, lyophilized, and then stored, protected from light, at $-20°C$. To administer the hormone, an aliquot is resuspended in 1 ml of 0.9% NaCl to give a final concentration of 50 IU/ml; 0.1 ml is then injected into each animal.

The Time of Administration of the Gonadotropins

The times that the PMS and hCG are administered relative to each other and to the light cycle of the mouse room will affect both the developmental uniformity and the number of eggs that are recovered from superovulated female mice. For most strains, a 42- to 48-hour interval between the PMS injection and the hCG injection has been found to be optimal in terms of egg yield. Generally, ovulation takes place between 10 and 13 hours after injection of hCG, but to control the time of ovulation precisely it is important to administer the hCG prior to the release of endogenous LH. The time that endogenous LH is released in response to PMS is regulated by the light-dark cycle. Thus, animals purchased for superovulation from a commercial supplier should be given a few days to adjust to the light-dark cycle of the mouse room before the administration of the PMS. The time of endogenous LH release will vary depending on the strain, but a reasonable estimate for most strains is between 15 and 20 hours after the midpoint of the second dark period following injection of PMS (Gates 1971). For example, on a 5 am–7 pm light cycle, PMS is administered between 1 pm and 2 pm and the hCG 46–48 hours later, usually between noon and 1 pm, and thus at least 2–3 hours before endogenous LH release. (Note, that in some strains of mice, the amount of endogenous LH released in response to PMS in a prepubescent mouse may not be sufficient to induce ovulation [see Gates 1971]; in these cases the timing of the hCG injection will not be so critical.) After administration of hCG each female is placed in a cage with a stud male (one female and one male per cage) and checked for a copulation plug the next morning.

The Strain of Mice

Inbred and F_1 hybrid strains will vary with respect to the number of eggs that are produced after superovulation. Table 15 classifies a small subset of inbred and F_1 strains according to the ability of prepubescent females to respond to the administration of PMS and hCG. The strains fall into two categories: high responders, which ovulate between 40 and 60 eggs per mouse, and low responders, which ovulate 15 or fewer eggs per mouse.

Reproductive Performance of the Stud Males

To maximize the number of fertilized eggs recovered from a superovulated female, it is critical to use stud males with a good plugging performance and a high sperm count. Procedures for maintaining such stud males are discussed in Section B.

Table 15 Strain Survey of Ovulation Response to Exogenous Gonadotropins

High ovulators	Low ovulators
C57BL/6J	A/J
BALB/cByJ	C3H/HeJ
129/SvJ	BALB/cJ
CBA/CaJ	129/J
CBA/H-T6J	129/ReJ
SJL/J	DBA/2J
C58/J	C57/L
BALB/cByJ × C57BL/6JF1	BALB/CJ × A/JF1
C57BL/6J × CBA/CaJF1	C57BL/6J × DBA/2JF1

All females were prepubescent, on a 12-hr light-dark cycle. Doses of PMS and hCG were 5 IU/0.1 cc injected i.p.

The information in this table was provided by Michael P. Rosenberg, Mammalian Genetics Laboratory, Litton Bionetics, Inc., Basic Research Program, National Cancer Institute, Frederick Cancer Research Facility, Frederick, Maryland.

HOW TO INJECT A MOUSE INTRAPERITONEALLY

The essential steps are illustrated in Figure 32.

1. Pick the mouse up by the scruff of its neck as close to the ears as possible. Be sure to take up enough skin so that it cannot turn its head and bite you. Hold the tail by twisting it around your little finger.

2. Inject solution intraperitoneally with a size 16×0.5-mm needle, taking care to avoid either the diaphragm or the bladder. Wait briefly before withdrawing the needle so that liquid does not seep out. If by mistake the injection has been subcutaneous, a bleb will appear at the site of injection.

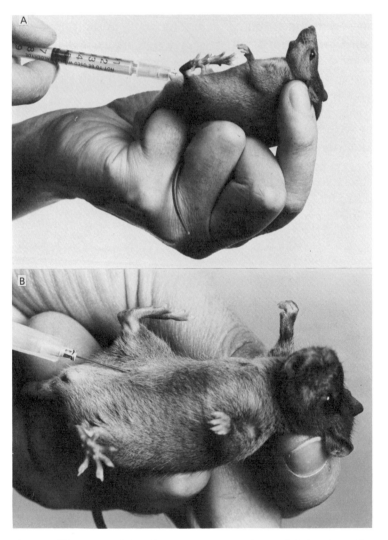

Figure 32 Method for humane intraperitoneal injection of a mouse.

MAKING PIPETS FOR COLLECTING AND HANDLING EMBRYOS

EQUIPMENT

Hard glass capillary tubing (1.5-mm o.d., BDH Chemicals Ltd., Poole,
 U.K.; Gallard-Schlessinger in U.S.A.) or 23-cm glass Pasteur pipets
Small bunsen burner or microflame
Tubing and mouthpieces
Glass test tubes, metal caps, and oven for sterilizing
Diamond pencil

PROCEDURE

1. Soften the glass by rotating the glass tubing or Pasteur pipet in the fine flame.
   If necessary, a microflame can be constructed as shown in Figure 33A.

2. Withdraw the tubing from the heat as quickly as possible and pull sharply to
   produce a tube with an internal diameter of about 200 μm.

3. For a neat break, score with a diamond pen and snap gently. Alternatively, pull
   on the cooled tubing until the two halves come apart.

4. Collect several pipets in a glass tube and, if required, sterilize by heating in an
   oven. A convenient holder for the pipets can be constructed from tubing and a
   mouthpiece, as shown in Figure 33B.

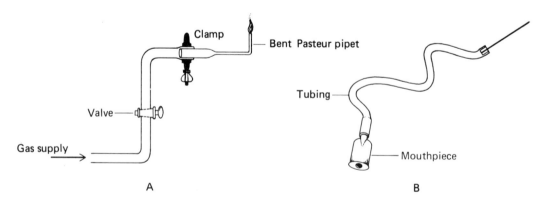

Figure 33 Apparatus for preparing pipets for handling and transferring embryos. On
the left, a microburner for pulling pipets; on the right, suggested arrangement for mouth
control of pipets.

RECOVERY OF PREIMPLANTATION EMBRYOS

Opening Up the Abdominal Cavity

EQUIPMENT

> 70% or 90% ethanol in a squeeze bottle
> Absorbent paper
> Scissors (one pair fine iris scissors, one pair regular surgical)
> Two pairs #5 watchmaker's forceps
> Alcohol burner or small bunsen burner if sterile techniques are to
>   be used (the scissors and forceps can be kept to one side in a
>   beaker containing cotton wool and ethanol)

PROCEDURE

1. Quickly place the mouse on top of the cage, so that it grips the bars firmly with its front paws. Then break its neck humanely by applying firm pressure at the base of the skull and sharply pinching and twisting between thumb and forefinger (Fig. 34) while at the same time pulling backward on the tail. The action is one of stretching rather than snapping and very little force is needed. Alternatively, a spatula or pencil can be used to apply pressure to the base of the skull.

2. Lay the animal on its back on absorbent paper and soak it thoroughly in 70–90% ethanol from a squeeze bottle; this important step reduces the risk of contaminating the dissection with mouse hair.

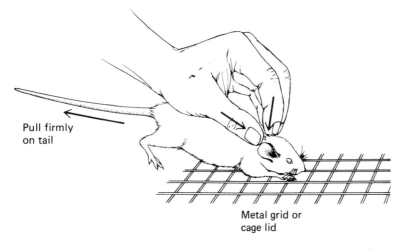

Pull firmly on tail

Metal grid or cage lid

Figure 34 Method for quick and humane killing of a mouse by cervical dislocation.

3. Pinch up the skin and make a small cut in the midline (the position is not critical) with the regular scissors (Fig. 35A). Holding the skin firmly above and below the cut, pull the skin toward the head and tail until the abdomen is completely exposed and the fur is well out of the way. Using forceps and fine scissors cut the body wall (peritoneum) as shown in Figure 35B. Push the coils of gut out of the way and locate the two horns of the uterus, the oviducts, and the ovaries (Fig. 35C).

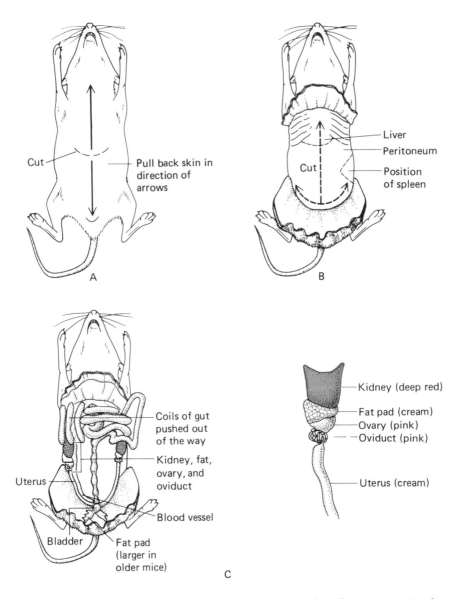

Figure 35 Dissection of reproductive organs of a female mouse. (A) The position of the small lateral cut in the skin is indicated by the dotted line. The skin is then pulled back in the direction of the solid arrows. (B) The peritoneum is cut in the direction of the dotted arrows. (C) Alimentary tract displaced to reveal reproductive organs in the floor of the body cavity.

*Collecting Fertilized Eggs and Removing Cumulus Cells with
Hyaluronidase*

EQUIPMENT

Pregnant females
— 35-mm petri dishes or embryological watch glasses
M2 medium (see Section G)
Hyaluronidase solution in M2 (see Section H)
Sterile transfer pipets
Stereomicroscope with understage illumination
    (preferably a ground-glass stage) with 20× and 40×
    magnification, e.g., Nikon SMZ-10 or Wild M5A
Fiber optic illumination (optional)

PROCEDURE

The fertilized eggs may be dissected out several hours before they are to be injected. When mice are maintained on a light-dark cycle with the midpoint of the dark period at midnight, it is convenient to begin dissecting out eggs at about 11 am to 12 noon. Later in the day, when the cumulus cells begin to fall off, the eggs are somewhat more difficult to recover.

1. Open up the abdominal cavity as described above. Grasp the upper end of one of the uterine horns with fine forceps and gently pull the uterus, oviduct, ovary, and fat pad taut and away from the body cavity. This will reveal a fine membrane (the mesometrium) which connects the reproductive tract to the body wall and carries a prominent blood vessel. Poke a hole in the membrane with the tips of a closed pair of fine iris scissors (Fig. 36).

2. Pull the oviduct, ovary, and fat pad taut, and cut between the oviduct and the ovary as shown (Fig. 36B).

3. The oviduct and attached segment of uterus can now be transferred to a 35-mm Petri dish or embryological watch glass containing M2 at room temperature (see Section G). Oviducts from several mice can be collected in the same dish.

4. Newly ovulated eggs, surrounded by cumulus cells, are found in the upper part of the oviduct or ampulla, which at this time is much enlarged. The fimbriated end of the oviduct, the infundibulum, is also swollen during ovulation and can easily be located under $20\times$ magnification (Fig. 36C). Transfer one oviduct at a time into another 35-mm petri dish containing M2 plus hyaluronidase (about $300\ \mu g/ml$) at room temperature and view though the stereomicroscope at $20\times$ or $40\times$ magnification. Using one watchmaker's forceps, grasp the oviduct next to the swollen infundibulum and hold it firmly on the bottom of the dish. Use the other forceps to tear the oviduct close to where the eggs are located, releasing the clutch of eggs. If the eggs do not flow out by themselves, use the forceps to push them out by gently squeezing the oviduct. If the eggs stick to the outside of the oviduct, allow the oviduct to sit for several minutes in the hyaluronidase. As the digestion removes the sticky cumulus cells, the eggs will be released. If the eggs stick to the forceps, simply lift the forceps out of the dish and the eggs will be retained by the surface tension and fall back to the bottom of the dish.

5. Allow the eggs to incubate in the enzyme for several minutes until the cumulus cells fall off. If necessary, pipet the eggs up and down a few times but do not leave the eggs in the hyaluronidase for more than a few minutes after the cumulus cells are shed, as this may be harmful. Using transfer pipets pick up the eggs and transfer them to a fresh dish of M2 to rinse off the hyaluronidase. Then transfer the eggs to M16 for culture at $37\,°C$ until needed (see Setting Up Microdrop Cultures).

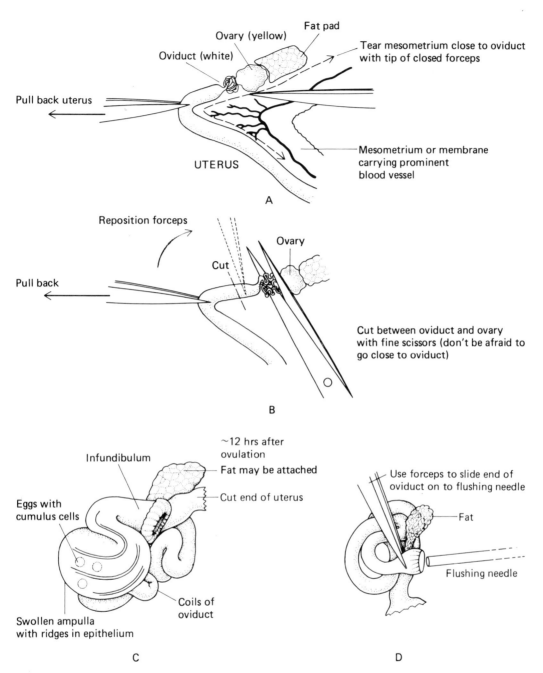

Figure 36 Collecting embryos from the oviduct. The ovary, oviduct, and end of the uterus are separated from the mesometrium (*A*) and a cut is made between the oviduct and the ovary (*B*). After repositioning the forceps, a second cut separates the oviduct from the uterus (*B*). If the oviduct is removed soon after fertilization, the eggs surrounded by follicle cells can be seen in the swollen ampulla and can be released by tearing the ampulla with fine forceps (*C*). At later stages the embryos are recovered by flushing the oviduct by inserting a needle into the end of the oviduct (infundibulum) (*D*).

Collecting Two- to Eight-cell Eggs and Eight-cell Compacted Morulae

Two- to eight- cell embryos are present in the oviduct 20–60 hours post coitum (p.c.). By this time the embryos have lost their cumulus cells and they can be flushed from the oviduct using about 0.1 ml of M2 in a 1-ml syringe attached to a 32-gauge hypodermic needle.

PROCEDURE

1. Insert the end of the needle into the opening of the oviduct, or infundibulum, which at this time is no longer swollen and must be located within the coils of the oviduct (Fig. 36D). To reduce the risk of tearing the oviduct, the end of the needle can be cut and ground to a blunt tip. Test that the syringe is free of air bubbles and the medium is flowing smoothly before inserting the needle. Hold the oviduct in place with fine forceps. If the embryos are to be cultured, the needle, forceps, etc., should be sterile.

    By 60–72 hours p.c. some compacted morulae will have entered the uterus, so for efficient recovery both the oviduct and uterus must be flushed. Cut the uterus above the cervix, as described for the collection of blastocysts (below).

2. If flushing proves too difficult, there are alternative ways of collecting the embryos. For example, the oviduct can simply be torn at several points along its length with fine forceps. More elegantly, "stroke" the oviduct with the blunt edge of a needle toward the uterine end and squeeze out the embryos.

SETTING UP MICRODROP CULTURES

EQUIPMENT

35-mm sterile tissue culture dishes
Light paraffin oil (Section H)
M16 medium (Section G)
Humidified 37°C incubator with an atmosphere of 5% CO_2 and 95%
   air, or 5% CO_2, 5% O_2, and 90% N_2
Sterile transfer pipets

PROCEDURE

1. Dispense 20- to 40-μl drops of medium in an array on the bottom of the tissue culture dish and then flood the dish with paraffin oil (Fig. 37).

2. Transfer the embryos into the drops and place in the incubator. The drops should be set up a short time in advance to allow for temperature and CO_2 equilibration.

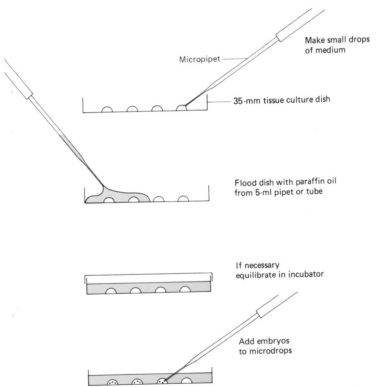

Figure 37 Setting up microdrops for embryo culture. Small drops of medium are formed on a 35-mm plastic tissue culture dish. The dish is flooded with paraffin oil and placed in a 37°C gassed incubator to equilibrate.

COLLECTING BLASTOCYSTS

Blastocysts can be flushed from the uterus between 3.5 and 4.5 days p.c.

PROCEDURE

1. To remove the uterus at this stage, grasp it with forceps just above the cervix (located behind the bladder) (Fig. 38A) and cut across with fine scissors (Fig. 38B). Pull the uterus upward to stretch the mesometrium and trim this membrane away close to the wall of the uterine horns (Fig. 38C). Then cut the uterus below the junction with the oviduct. The utero-tubal junction acts as a valve, and if the cut is made across the oviduct rather than the uterus, flushing will be very difficult.

2. Place the uterus into M2 and flush both horns with about 0.2 ml of medium using a 25-gauge needle and a 1- or 2-ml syringe (Fig. 38D). This can be done without using a dissecting microscope.

   The yield of blastocysts is very low after they have hatched from the zona and have attached to the uterus. The yield can sometimes be increased by leaving the utero-tubal junction intact, blowing up the uterine horns by injecting medium, and then cutting the junction to release the fluid.

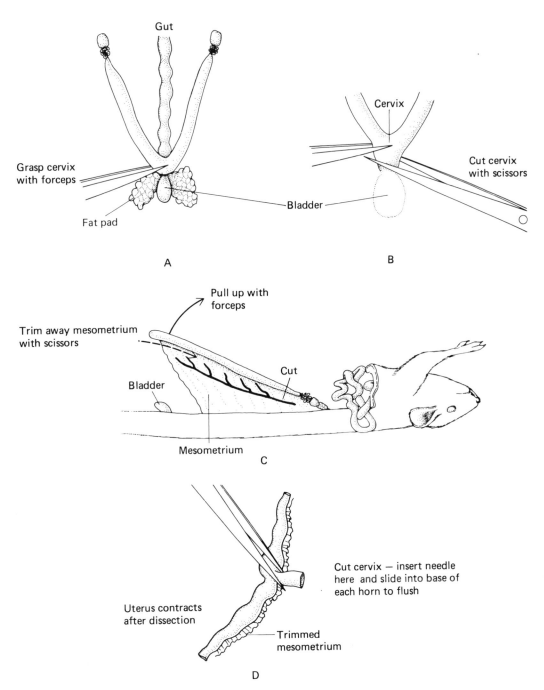

Figure 38 Dissection for flushing embryos from the uterus. The uterus is removed by cutting across the cervix (*A* and *B*) and the membrane (mesometrium) holding the uterus to the body wall (*C*). Be sure to cut the uterus below the junction with the oviduct (*C*). Place the uterus in a small volume of medium, and flush out each horn with about 0.2 ml of medium (*D*).

REMOVING THE ZONA PELLUCIDA

The zona pellucida can be removed either by exposing the embryos to low pH (pH 2.5) or by digestion with Pronase. The second method is more controllable and can be used on fixed embryos, provided that they are washed free of any buffer that will inhibit the enzyme (e.g., cacodylate). The first method has the advantage of speed, but cannot be used on fixed material.

EQUIPMENT

  35-mm tissue culture dishes
  M2 or M16 medium
  Acidic tyrode solution (Section H) or
  Pronase (0.5% in M2, Section H)
  Transfer pipets
  Stereomicroscope

PROCEDURE

Method Using Acid Tyrode

1. Transfer the embryos in as small a volume as possible (not more than a few microliters) to about 1 ml of acid tyrode at room temperature in a 35-mm tissue culture dish. Mix very gently to break down any local concentration of medium but without dispersing the embryos.

2. Observe continuously under the stereomicroscope. As soon as the zona is dissolved, the embryos will stick to the dish. Immediately flood the dish with excess medium.

3. Collect the embryos and transfer back to microdrop culture.

Method Using Pronase

1. Transfer the embryos to microdrops containing Pronase solution.

2. Incubate at 37°C until the zona is removed (3–10 min).

3. Wash thoroughly by passage through several drops of M2 and incubate as before.

IN VITRO FERTILIZATION

One application of this technique* is to obtain large numbers of early-cleavage-stage embryos that are developing more synchronously than those collected after natural mating. Another use is to generate offspring from mice that, for one reason or another, will not mate or carry litters to term. In this case, the in vitro-fertilized eggs are transferred to a pseudopregnant foster mother as described in Section C, Oviduct Transfers.

The in vitro fertilization technique involves superovulating females and fertilizing the eggs so obtained with sperm taken from the epididymis of the male. This has been found to give more synchronous development than sperm from the vas deferens (H. Pratt, unpubl.).

PROCEDURE

1. Inject females with PMS and hCG as for superovulation (Section C, Superovulation).
2. Kill males 12 hours after the females have been injected with hCG. (Do not use males that have mated within the previous 3 days or have abstained for more than a week.) Dissect out the epididymis, removing as much excess fat as possible. Cut the vas deferens close to the epididymis and place the latter into a 500-μl drop of pregassed Whittingham's medium (+30 mg/ml BSA) under oil (see Section G for Whittingham's medium formulation). Gently squeeze out the sperm using watchmaker's forceps. The sperm should then be incubated for 1.5 hours at 37°C to capacitate them. This step is essential.

3. Kill females 12.5 hours post-hCG injection and dissect out the oviducts. Release up to 10 cumulus masses into a 1000-μl drop of pregassed Whittingham's medium (+30 mg/ml BSA) under oil. Incubate at 37°C in 5% CO_2, 95% air.

4. At 13.5 hours post-hCG add 100 μl of sperm to the drops containing eggs. This gives a sperm concentration in the range 1×10^6 to 2×10^6 sperm/ml.

5. Incubate the eggs and sperm for 4 hours at 37°C.

6. Transfer the eggs into pregassed sterile M16 (+4 mg/ml BSA), taking as few sperm as possible.

COMMENTS

1. The time between PMS and hCG injections must not be greater than 48 hours and it is best for the time to be exactly 48 hours (though 46 hours is acceptable).

*Information provided by Dr. Hester Pratt, Department of Anatomy, University of Cambridge, Downing Site, Cambridge, U.K., based on method described by Fraser and Drury (1975).

2. If the sperm is difficult to squeeze out of the epididymis, it is best to discard it and use another male. Ideally, the sperm emerges very easily, like "toothpaste from a tube," and should need very little force.

3. Insemination earlier than 13.5 hours post-hCG does not result in fertilization.

4. If necessary, fertilized eggs may be removed from the medium containing the sperm as soon as 3 hours postinsemination.

5. When fertilized, the eggs show second polar body extrusion 3–6 hours postinsemination.

ETHANOL-INDUCED PARTHENOGENETIC ACTIVATION OF OOCYTES

Numerous studies have been initiated to investigate the influence of the maternal and paternal genomes on early mammalian development. For this type of study, parthenogenetic embryos provide a unique source of preimplantation and early postimplantation embryos that (by definition) develop in the absence of any contribution from a male gamete. Much interest has also been expressed in utilizing the EK cell lines that can be established directly from parthenogenetic blastocysts (Kaufman et al. 1983). For example, if the eggs carry recessive mutations or are mutagenized after fertilization, the EK cells will become homozygous for the mutation when the cells diploidize.

The technique for activating oocytes with ethanol* is described in several references, e.g., Kaufman (1978a,b).

PROCEDURE

1. Mice 8-12 weeks old are superovulated with 5 IU of PMS followed 48 hours later by 5 IU of hCG (see Superovulation). Kill animals 17 hours after the hCG and recover eggs surrounded by cumulus cells (see Recovery of Preimplantation Embryos).

2. Release the eggs with their cumulus cells into freshly prepared 7% ethanol (analytical reagent grade) in Dulbecco's phosphate-buffered saline (PBS) in a 3-cm tissue culture dish and leave for 5 minutes. Wash through three changes of PBS and two changes of M2 in 3-cm dishes. The cumulus masses are then transferred individually into drops of M16 under paraffin oil and incubated in a 37°C, humidified, 5% CO_2 incubator.

3. After 5 hours remove the cumulus cells by hyaluronidase treatment (see Collecting Fertilized Eggs for Microinjection and Removing the Cumulus Cells with Hyaluronidase) and classify the activated oocytes under phase-contrast or Nomarski optics (for classification of oocytes, see Fig. 39, also Fig. 11). Separate the different classes into drops of M16 under oil.

Eggs derived from certain F_1 hybrid females (e.g., C57BL/6 × CBA) can be cultured from the one-cell stage. Eggs from other sources, e.g., inbred strains, suffer a two-cell block and will not proceed beyond the first mitotic division when retained in culture. Further development of these oocytes can be obtained by transferring them to the oviducts of 0.5-day p.c. pseudopregnant recipient females.

*Information supplied by Dr. M.H. Kaufman, Department of Anatomy, University of Cambridge, U.K.

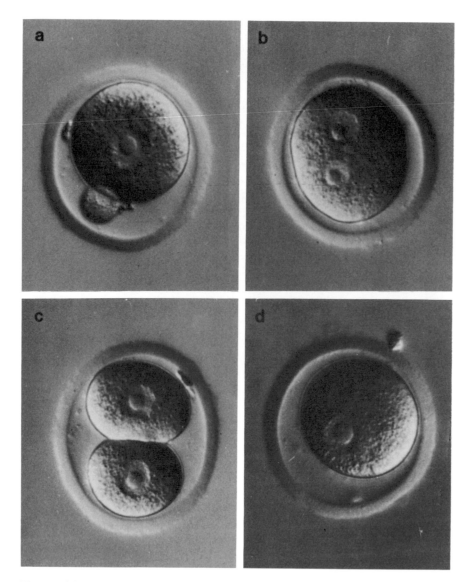

Figure 39 The four classes of parthenogenetic eggs that can be distinguished at 4–6 hr after ethanol-induced activation. (a) Single pronuclear haploid egg with extruded second polar body (uniform haploid). (b) Two-pronuclear presumptive diploid egg (heterozygous diploid). (c) Immediate cleavage embryo with two approximately equal-sized blastomeres (mosaic haploid). (d) Single pronuclear diploid egg (heterozygous diploid).

The activation rate is approximately 80–85% of oocytes exposed to alcohol. Normally the highest proportion of the activated population consists of haploid parthenogenones that develop a single haploid pronucleus following extrusion of the second polar body. Diploidization of this class of oocyte can be achieved by suppression of the formation of the second polar body by incubation in cytochalasin D-containing medium (Kaufman 1978a,b).

DISAGGREGATION OF CLEAVAGE-STAGE EMBRYOS INTO INDIVIDUAL CELLS

EQUIPMENT

Embryos that have been allowed to recover for about 1 hour after removing the zonae

Calcium-free M16 containing 6 mg/ml BSA rather than the usual 4 mg/ml M16 with 4 mg/ml BSA (Section G)

Bacteriological plastic culture dishes

This procedure has been described by Ziomek and Johnson (1980).

PROCEDURE

1. Incubate the embryos without zonae in calcium-free medium for 10–15 minutes in drops under oil at 37 °C in a humidified incubator with 5% CO_2. Disaggregate the blastomeres by pipeting through a flame-polished glass pipet.

2. Remove blastomeres from the calcium-free medium as soon as possible. The isolated blastomeres are very sticky, so place them individually or in small groups in single drops of conventional M16 medium in petri dishes to which they will not adhere (e.g., bacteriological dishes made by Sterilin). Cell death may be reduced by transferring the embryos into calcium-containing medium before disaggregation by pipeting (information provided by C.F. Graham).

MAKING AGGREGATION CHIMERAS

EQUIPMENT

Precompaction cleavage stage embryos (e.g., 4–8 cell)
Pronase
M2
Microdrops of M16 in tissue culture dishes

1. Remove the zonae with Pronase. Observe the embryos throughout the incubation and take them out as soon as the zona distorts and swells (3–10 min). Wash in M2 and transfer to M16.

2. Incubate at 37°C until the zonae have completely gone (5–30 min). (If the zona is still present, the embryos will not mingle.)

3. Put the embryos together in pairs or triplets in microdrops in M16. Then push the embryos together with a glass pipet or needle and handle the dish gently so that contact is not broken. After 1 hour of incubation at 37°C the blastomeres should have mingled into a single embryo. If not, nudge them together again. Continue incubation until the embryo has compacted (1 day of culture), or until it has developed into a blastocyst (2 days of culture).

4. At either stage, the chimera is transferred to the uterus of a 2.5-day p.c. pseudopregnant female.

Blastocysts have a better chance of implantation, but balanced against this is the risk of poor development in vitro if culture conditions are not ideal. If only a small number of chimeras are to be transferred, use a genetically different 2.5-day p.c. pregnant female so that the litter size is large enough (five to seven).

ISOLATION OF THE INNER CELL MASS BY IMMUNOSURGERY

This technique was first described by Solter and Knowles (1975) for selectively killing the outer trophectoderm cells (TE) of the blastocyst. The principle is outlined in Figure 40.

PROCEDURE

1. Preparation of sera. Rabbit anti-mouse serum is prepared by bleeding a rabbit 10 days after three fortnightly injections of 4×10^8 mouse cells (for example, spleen cells obtained from any strain of mouse, or mouse tissue culture cells). Allow the blood to clot and collect serum. Heat-inactivate at 56°C for 30 minutes to destroy complement and store in aliquots at -70°C.

   For guinea pig complement, it is sufficient to use guinea pig serum rather than purified complement. The serum can either be purchased (e.g., from Wellcome Laboratories or Gibco, as freeze-dried guinea pig serum) or prepared from fresh guinea pig blood obtained by cardiac puncture. However, the source of guinea pigs appears to be important. Animals kept in specific pathogen-free (SPF) conditions are best since they are less likely to have developed natural antibodies against bacterial carbohydrates which may cross-react with carbohydrates on the surface of mouse cells. If the serum batch is toxic to mouse cells, it is sometimes (but not always) possible to remove the endogenous antibodies by preadsorbing the serum with agar (Cohen and Schlesinger 1970).

2. Dilute serum 1:3 with Dulbecco's modified Eagle's medium (DMEM) (or similar medium) to 3 ml, add 80 mg of agar (Difco Noble agar) or agarose (Sigma or Calbiochem), leave on ice 30–60 minutes with occasional shaking, and centrifuge at 4°C to pellet agarose. Store supernatant in aliquots at -70°C. This serum can usually be used without further dilution.

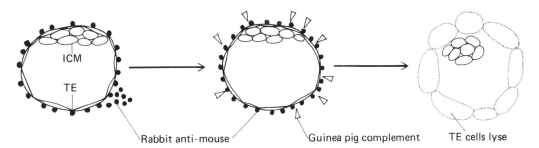

Figure 40 Schematic representation of the technique of immunosurgery for isolating inner cell masses. Blastocysts are incubated with rabbit anti-mouse serum (●), washed thoroughly, and exposed to guinea pig complement (▽). Only the outer trophectoderm cells are lysed; the inner cell mass cells are protected from exposure to the rabbit antibodies by the tight permeability seal of the trophectoderm.

3. Transfer blastocysts in a few microliters of medium (e.g., M2 or 20 mM HEPES-buffered DMEM + 10% heat-inactivated fetal bovine serum) to a much larger volume (e.g., 3 ml) of DMEM containing rabbit anti-mouse serum diluted 1:30 to 1:100 (the optimal dilution should be determined in preliminary tests). Incubate 37°C for 10 minutes. Antibodies bind to the outer trophoectoderm cells but are prevented by the zonular tight junctions between these cells from penetrating into the blastocoel and binding to the inner cell mass cells.

4. Wash the blastocysts in two changes of DMEM supplemented with HEPES and heat-inactivated fetal bovine serum.

5. Incubate with guinea pig complement (diluted 1:3 to 1:50) at 37°C for 30 min. The outer trophectoderm cells should begin to lyse within a few minutes.

6. Wash the inner cell masses and transfer them to culture medium (see Section G). The dead trophectoderm cells can be removed by gentle pipeting in a finely drawn Pasteur pipet, but this removal is not necessary for subsequent development.

   The in vitro growth and differentiation of isolated inner cell masses has been described (Solter and Knowles 1975; Hogan and Tilly 1978a,b; Wiley et al. 1978). The immunosurgery process can be repeated on inner cell masses that have differentiated a continuous outer layer of endoderm (Hogan and Tilly 1977).

ISOLATING POSTIMPLANTATION-STAGE EMBRYOS

Postimplantation embryos require more complex media than early cleavage stages if the cells are to remain viable and healthy during the dissection. It is therefore advisable to dissect the embryos into Dulbecco's modified Eagle's medium with 10% calf or fetal bovine serum. Another important function of the serum is to stop the tissues from being too sticky. HEPES buffer (20 mM, pH 7.4) is added to maintain the pH during handling outside the incubator. Other complex media would be satisfactory and Alpha medium is also frequently used.

Early Egg Cylinder Stage (∼5.5 days p.c.)

PROCEDURE

1. Cut one horn of the uterus below the oviduct and grasp the end firmly with forceps. Pull upwards and separate the uterus from the mesometrium using the tips of forceps or scissors (Fig. 41A). Then pull the uterus down tautly and out to one side.

2. Insert the tips of a fine pair of scissors (Fig. 41B) into the antimesometrial wall of the uterus near the cut end. Pointing the scissors slightly upwards, "slide" them along, at the same time carefully cutting through the uterus wall. The decidua can now be "shelled" out of the uterus with forceps (Fig. 41C) and transferred to a dish of medium. At this stage the deciduum does not always separate cleanly from the endometrial stroma and is rather ragged.

3. Tease the deciduum apart with fine forceps (the two walls of the uterus and their epithelia have not yet fused completely and there is a distinct groove, see Fig. 14) and look for the embryo as a dark speck or streak. The embryo can be shelled out with the tips of the forceps. The dark tissue is the trophoblast, whereas the egg cylinder is very fragile and transparent (Fig. 41D).

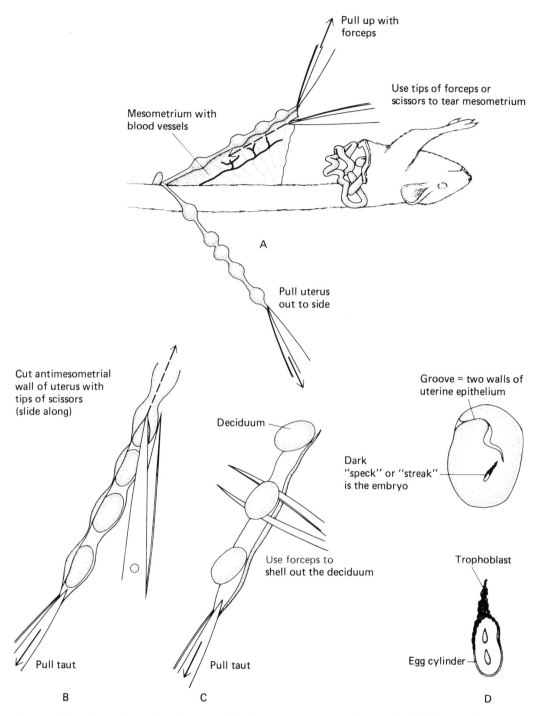

Pull up with
forceps

Use tips of forceps or
scissors to tear mesometrium

Mesometrium with
blood vessels

A

Pull uterus
out to side

Cut antimesometrial
wall of uterus with
tips of scissors
(slide along)

Deciduum

Groove = two walls of
uterine epithelium

Dark
"speck" or "streak"
is the embryo

Use forceps to
shell out the deciduum

Trophoblast

Egg cylinder

Pull taut   Pull taut

B   C   D

Figure 41   Dissection of early egg cylinder-stage mouse embryos (∼5.5 days p.c.).

Egg Cylinder Stage (~6.5 days p.c.)

PROCEDURE

The decidua can be dissected from the uterus as described above. Alternatively:

1. Remove the uterus intact by cutting across the cervix and the two utero-tubal junctions, and place in a dish containing medium. Cut into individual swellings as shown in Figure 42A. The muscle layer is removed with forceps as shown in Figure 42B.

2. Dissect the decidual tissue (Fig. 42C) to reveal the small embryo, which can be shelled out with the tip of a closed forceps. The embryo can be further dissected into ectoplacental cone (trophoblast) and egg cylinder using tungsten needles in a scissor-like action. For details of making needles, see Section H. The techniques are described by Snow (1978b).

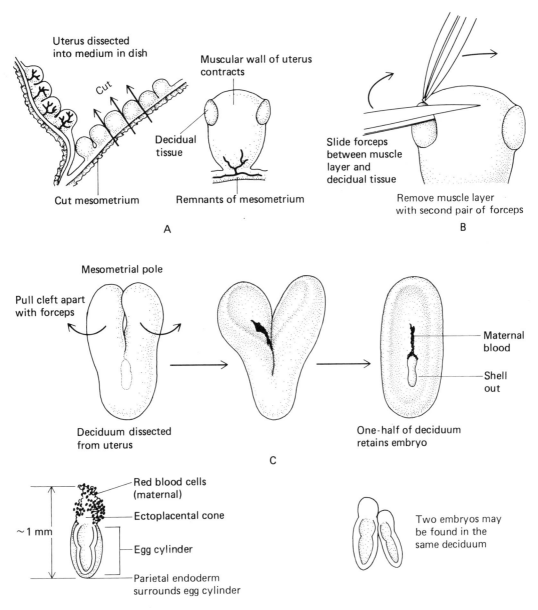

Uterus dissected
into medium in dish

Cut

Muscular wall of uterus
contracts

Decidual
tissue

Cut mesometrium

Remnants of mesometrium

A

Slide forceps
between muscle
layer and
decidual tissue

Remove muscle layer
with second pair of forceps

B

Mesometrial pole

Pull cleft apart
with forceps

Maternal
blood

Shell
out

Deciduum dissected
from uterus

One-half of deciduum
retains embryo

C

Red blood cells
(maternal)

Ectoplacental cone

~ 1 mm

Egg cylinder

Parietal endoderm
surrounds egg cylinder

Two embryos may
be found in the
same deciduum

Figure 42 Dissection of egg cylinder-stage mouse embryos (∼6.5 days p.c.).

Early primitive streak (∼7.5 days p.c.)

PROCEDURE

1. Dissect the decidua and embryos, essentially as above (Fig. 43A,B).

At this stage, and earlier, the egg cylinder can be separated into ectoderm, endoderm, and mesoderm by a combination of enzymatic digestion and mechanical dissection (see Separation of Tissue Layers Separated by a Basement Membrane, and Snow 1978b.)

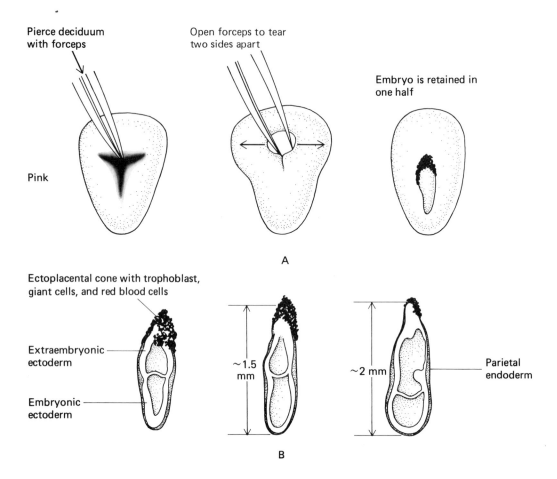

Figure 43 Dissection of early primitive streak-stage mouse embryos (∼7.5 days p.c.). (*A*) Technique for separating the decidual tissue to expose the embryo, which can then be flipped out with the tips of closed forceps. (*B*) The embryos are not synchronized in their development; more advanced ones have a small allantois (*far right*).

Early neural fold (∼8 days p.c.)

See Figure 44A and B.

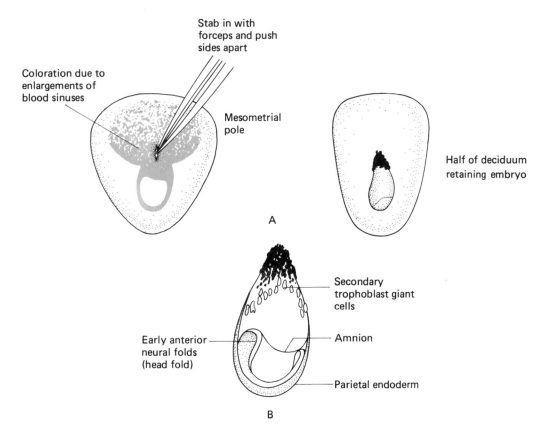

Figure 44 Dissection of early neural fold-stage mouse embryo (∼8 days p.c.). (*A*) The technique for separating embryo from decidual tissue is essentially as in Fig. 43. (*B*) The embryo now has a distinctive ∼ shape with large anterior neural folds and a deep neural groove.

Early Somite (~8.5 days p.c.)

PROCEDURE

For dissection at this stage, the easiest method is:

1. Cut off the mesometrial one-third of the deciduum using forceps, as shown in Figure 45A.

2. Pull the two halves of the deciduum apart with forceps and gently shell out the embryo.

3. In most embryos, the allantois has not yet fused with the chorion (Fig. 45B). More advanced embryos will have begun "turning" and in a few the heart will be beating.

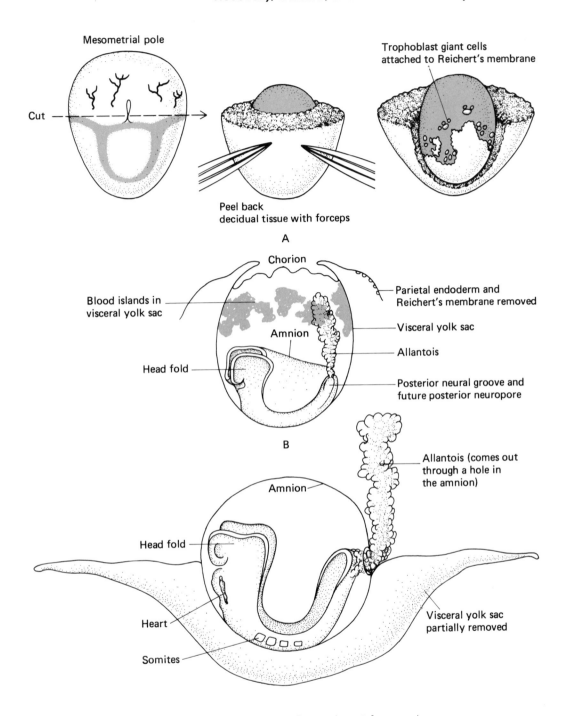

Figure 45 Dissection of early somite-stage embryos (∼8.5 days p.c.).

ISOLATING EXTRAEMBRYONIC MEMBRANES

The best source of trophoblast tissue free of maternal contamination is the 7.5- to 8.5-day p.c. ectoplacental cone (Rossant and Croy 1985). For isolating the parietal yolk sac, visceral yolk sac, and amnion simultaneously a good age to use is 13.5 days p.c. In older embryos the parietal yolk sac begins to degenerate and has disappeared by day 15–16. The isolation of membranes from the 13.5-day embryo is described below.

PROCEDURE

1. Remove the uterus and cut transversely into individual decidua (Fig. 42A). Remove the muscle layer of the uterus (which will spontaneously retract in some cases (Fig. 46A). Rotating the embryo, and using the tips of two pairs of forceps in a "nipping" action, cut around the junction of Reichert's membrane with the placenta (Fig. 46B). The membrane with parietal endoderm cells attached then can be removed and transferred to fresh medium. Some trophoblast cells will be attached, the quantity depending on the age and strain of the mouse. (C3H/ He mice have very few at 13.5 days p.c.) By 15 days, Reichert's membrane is very thin and most of the parietal endoderm cells have degenerated.

2. For observing parietal endoderm cells in situ, Reichert's membranes can be attached onto the surface of a tissue culture dish. Stab around the periphery of each membrane with the closed tips of a pair of forceps, thereby anchoring the membrane into the plastic (experiment with several brands of dish to find the one with the softest surface). Rinse the membrane gently in serum-free medium and, if required, fix with a 1:1 mixture of methanol/acetone for 5 minutes at room temperature, air-dry, and store at −70°C (Lane et al. 1983). (A 1:1 mixture of methanol and acetone will not dissolve the plastic dish.)

3. After removing the parietal endoderm, cut off the placenta and upper part of the visceral yolk sac (VYS). The VYS and embryo (surrounded by the amnion) can now be separated (Fig. 46C). Note that clumps of visceral endoderm cells are easily broken off from the convoluted region nearest to the placenta; if care is not taken these can significantly contaminate the amnion or parietal yolk sacs dissected into the same dish. (For discussion of the problem of cross-contamination, see Dziadek and Andrews 1983.)

Toward the end of gestation the amnion becomes very "slimy" and difficult to handle. The copious glycosaminoglycan secretion can be removed by incubating the membrane briefly in hyaluronidase (∼300 μg/ml in DMEM).

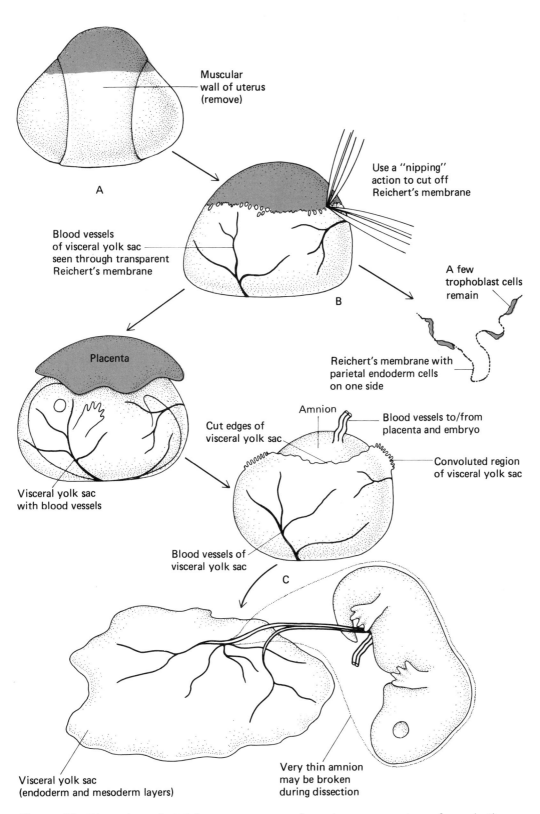

Muscular wall of uterus (remove)

A

Use a "nipping" action to cut off Reichert's membrane

Blood vessels of visceral yolk sac seen through transparent Reichert's membrane

B

A few trophoblast cells remain

Reichert's membrane with parietal endoderm cells on one side

Placenta

Visceral yolk sac with blood vessels

Amnion

Cut edges of visceral yolk sac

Blood vessels to/from placenta and embryo

Convoluted region of visceral yolk sac

Blood vessels of visceral yolk sac

C

Visceral yolk sac (endoderm and mesoderm layers)

Very thin amnion may be broken during dissection

Figure 46 Dissection of 13.5-day p.c. mouse embryo to recover extraembryonic tissues (parietal endoderm on Reichert's membrane, visceral yolk sac, and placenta).

SEPARATION OF TISSUE LAYERS SEPARATED BY A BASEMENT MEMBRANE

The two methods described here can be used to separate either the endoderm and ectoderm of the egg cylinder, or the endoderm and mesoderm layers of the visceral yolk sac.

PROCEDURE

1. Enzymatic method (pancreatin-trypsin). The idea is to incubate the tissues briefly at 4°C with a mixture of crude proteases (pancreatin and trypsin, see Section H, Solutions), followed by a period of recovery at 37°C before mechanical dissection (Levak-Svajger et al. 1969; Dziadek and Adamson 1978; Hogan and Tilly 1981). It is not clear how this treatment works and it is possible that in response to the trauma of exposure to exogenous proteases the cells also produce their own proteases for degrading the basal lamina between the epithelial and mesenchymal (or endodermal and ectodermal) layers.

   a. Endoderm, ectoderm, and mesoderm of 6.5- to 7.5-day p.c. egg cylinder. Serum contains protease inhibitors, so first rinse the dissected egg cylinders in medium (e.g., DMEM) without serum. Then incubate at 4°C in a small volume of pancreatin/trypsin solution (Section H, Solutions). Although some studies have used incubation times of up to 1 hour (Dziadek and Adamson 1978), 5 minutes were found to be adequate by others (Hogan and Tilly 1981). Flood the dish with DMEM containing 10% serum, transfer to fresh medium with serum, and incubate at 37°C for 1 hour to allow the cells to recover. Before mesoderm formation the tissue layers may be dissected manually using tungsten needles (see Section H, Supplies) or separated by drawing, cut end first, into a glass pipet pulled out to have a bore slightly narrower than the diameter of the egg cylinder. In older egg cylinders the mesoderm remains with the ectoderm and can be dissected with tungsten needles. Details of these procedures can be found in Snow (1978b).

   b. Separating endoderm and mesoderm of the visceral yolk sac. The epithelial layer of visceral endoderm can be separated from the underlying mesoderm (endothelial cells, blood islands, fibroblasts) after incubating visceral yolk sac (from a 10.5- to 13.5-day embryo) in the pancreatin/trypsin solution (Section H, Solutions); 0.1–1.5 hours at 4°C is usually required. Transfer the tissue to DMEM with 10% fetal calf serum and tease the two layers apart with watchmaker's forceps. The separate layers can be further dissociated into single cells by incubation in trypsin/EDTA (Section H, Solutions).

2. Nonenzymatic procedure (1 M glycine). A useful nonenzymatic procedure applicable to cylinders or visceral yolk sacs has been described (Dziadek 1981). Tissues are rinsed in PBSA and transferred to 1 M glycine, 2×10^{-3} M EDTA (pH 7.3). A 3- to 5-minute incubation at room temperature or at 37°C is sufficient for egg cylinders, while 30–60 minutes are needed for visceral yolk sacs. As the incubation time increases the endoderm cells are more likely to dissociate into single cells rather than as sheets.

ISOLATION AND CULTURE OF GERM CELLS FROM THE GENITAL RIDGE

This procedure* is for the isolation of germ cells from the genital ridges of fetal mice from 11.5 days p.c. onwards. Since different embryos from the same mother can vary in their stage of development, it is advisable to classify them according to the morphology of the hind limb bud, as shown in Figure 47. Male and female genital ridges cannot be distinguished until stage 6. The aim of the EDTA treatment is to dissociate the germ cells loosely from the stroma so that they can be released with minimal contamination with somatic cells.

*Information provided by Dr. Anne McLaren, MRC Mammalian Development Unit, London NW1 2HE, U.K.

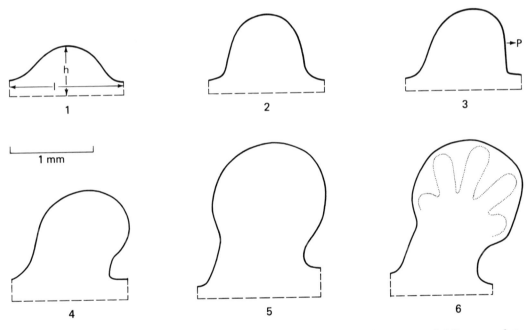

Figure 47 Morphology of hind limb bud from 10.5–12.5 p.c. fetuses useful for correlating the differentiation of the genital ridges. Male and female genital ridges cannot be distinguished until stage 6. (*1*) Length (l) > height (h); (*2*) h > l; (*3*) the limb bud is asymmetric, posterior (p) reentrant; (*4*) posterior and anterior reentrant; (*5*) limb bud is symmetric and has a circular outline; (*6*) the limb bud has an angulated outline and rays of toes are just visible.

EQUIPMENT

> Ophthalmic suture needles (#6 half-curved) (see Section H, Supplies)
>
> Needle Holder (see Section H, Supplies)
>
> Culture medium (DMEM or similar) with 10% fetal bovine serum
>
> Dissection medium—phosphate-buffered saline complete (Section H), or M2 (Section G)
>
> EDTA in phosphate-buffered saline plus glucose (see Section H, Solutions)
>
> Fine scissors and watchmaker's forceps
>
> 35-mm plastic petri dishes or glass embryological watch glasses
>
> Stereomicroscope

PROCEDURE

For long-term culture, instruments should be sterilized and antibiotics added to the medium.

1. Dissect the embryos from their membranes and place on an absorbent surface. With scissors, cut off the anterior half of the embryo just below the armpits (Fig. 48A). Make a cut along the ventral midline of the posterior half and scoop out the liver and intestines with the tips of the scissors (Fig. 48B).

2. Transfer the embryo fragment to a dish of dissection medium and turn it onto its back. Hold the embryo with forceps and remove remnants of intestines, etc., with the curved needle (Fig. 48C). The genital ridges lie on the dorsal wall of the embryo, adjacent to the shield-like mesonephros. Slide the needle behind the genital ridge and mesonephros and cut them from the embryo.

3. Transfer genital ridge and mesonephros to fresh medium with forceps. Using the curved needle, cut the mesonephros from the genital ridge (Fig. 49). From about 12.5 day p.c. (stage 6), the ovaries and testes can be distinguished by their morphology. Testes are "striped" and larger than ovaries at the same stage, while the ovaries are "spotted" and smaller than the adjacent mesonephric shield.

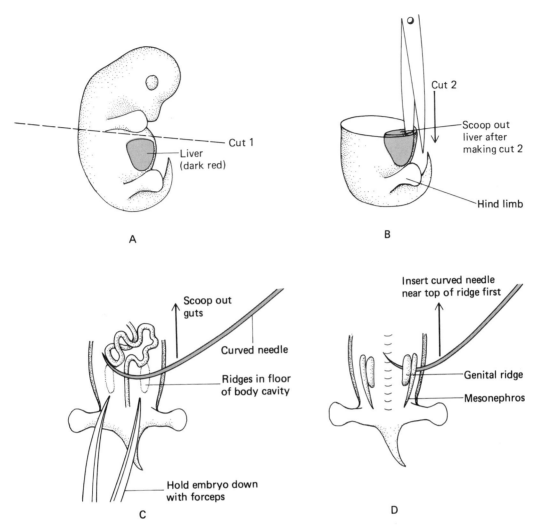

A

B

C

D

Figure 48 Dissection of genital ridges and mesonephros from a 12.5-day p.c. mouse embryo.

4. Transfer the genital ridges to EDTA/saline for about 15 minutes at room temperature. Return the ridges to the dissection medium. Holding each ridge with forceps, puncture the ridge with sharp stabs of a 26-gauge hypodermic needle (Fig. 49). This releases the germ cells but few of the stromal cells. About 20 stabs are sufficient; too-vigorous puncturing releases more germ cells but also increases the contamination with stromal cells. The germ cells can be recognized by their morphology under phase-contrast microscopy; they are large and have a smooth, often blebby outline (De Felici and McLaren 1983). The same reference describes details of a culture system for these germ cells.

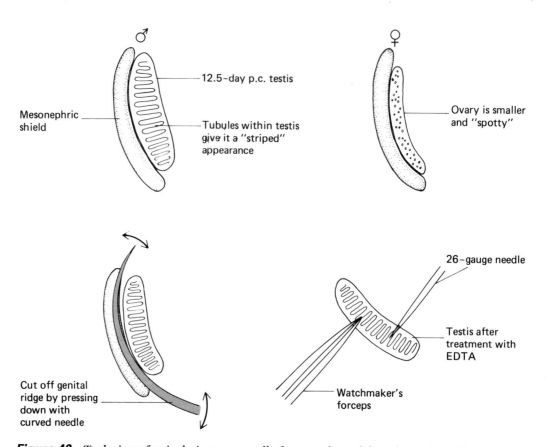

Figure 49   Technique for isolating germ cells from male and female genital ridges.

ISOLATION OF OVARIAN OOCYTES AND CULTURE OF FOLLICLE CELLS

Several methods have been used to isolate oocytes at different stages of maturity from the ovaries of juvenile or adult mice (e.g., Schultz et al. 1979; Sanford et al. 1984). The principle is to dissect out the ovaries, remove fat and connective tissue manually, and then puncture the tissue under the dissecting microscope using fine needles. The released oocytes are collected with a mouth-operated pipet (see Making Pipet for Collecting and Handling Embryos). When ovaries from immature mice (3–10 days after birth) are used, it is first necessary to loosen the stromal cells and oocytes by incubating the ovaries at 37°C in buffered saline, either without Ca^{++} or Mg^{++}, or with collagenase or trypsin (for details, see references).

Techniques have also been described for culturing follicle (granulosa) cells from various mammalian species (for review, see Hsueh et al. 1984). The following method has been used successfully for mouse (B.L.M.H.). Superovulate immature females as described (Superovulation). Then 9–10 hours after hCG (i.e., before ovulation), remove ovaries under sterile conditions into a tissue culture dish containing DMEM plus 10% fetal bovine serum. Holding the ovary with fine forceps, stab at the swollen, mature follicles with a sterile 26-gauge needle. The follicle cells are released in thick clumps which can be dissociated by gentle pipeting. Transfer cells to fresh medium and incubate at 37°C. The cells attach and spread and become very flattened. The cells from four mice (8 ovaries) will form a confluent monolayer on a 9-cm dish.

```
WEEK TWO :    VASECTOMY OF MALE MICE
```

Vasectomized males are needed to produce pseudopregnant females for oviduct and uterine transfers. Females in estrus should be selected as usual. After mating with a sterile male, the female reproductive tract becomes receptive for transferred embryos, even though her own unfertilized eggs degenerate. A pseudopregnant female will not resume her natural cycle for about 11 days. As an alternative to producing vasectomized males, sterile males of the genotype *T/t* can be used. For oviduct transfers, use 0.5-day p.c. pseudopregnant females, i.e., females mated the night before. For uterine transfers use 2.5-day (or 3.5-day) p.c. females.

EQUIPMENT

Avertin (anesthetic) (see Section H)
Animal balance
1-cc syringe
26-gauge, 1/2-in. needle
Animal clippers (optional)
70% ethanol
Tissues
Dissection scissors
Two pairs #5 watchmaker's forceps, sharpened
One pair blunt forceps
Surgical silk suture (size 5-0)
Surgical needle
Autoclips and applicator (optional)
Alcohol burner
Fiber optic illuminator

PROCEDURE

1. Anesthetize a male mouse. Males at least 2 months of age, from any strain with a good breeding performance, can be used. Weigh a mouse and inject it intraperitoneally with 0.015–0.017 ml of 2.5% Avertin per gram of body weight. (See Section H for preparation of Avertin.)

2. Shave the abdomen. This can be omitted once expertise is achieved. Wipe with 70% ethanol.

3. Open the body wall. Sterilize all instruments by flaming with the alcohol burner. Cut the skin with the dissection scissors (a 1.5-cm transverse incision) at a point level with the top of the legs. Make a similar-sized transverse incision in the body wall; put one stitch through the body wall on one side of the incision, and leave a piece of silk suture in place. This helps to find the body wall later. Both testes can be reached through the one incision.

4. Tie off the vas deferens. Using the blunt forceps, pull out the fat pad on the left side (Fig. 50A). The left testis, vas deferens, and epidydimis will come with it. The vas deferens lies underneath the testis and can be recognized by a blood vessel running along one side. With a sharp forceps, poke a hole in the membrane beneath the vas deferens and pull through a loop of silk suture. Cut the loop to yield two pieces of suture, each 5–10 cm long. Tie a double knot with each thread (Fig. 50B). The knots should be 4–5 mm apart. Sever the vas deferens between the two knots. Pick up the fat pad with the blunt forceps and carefully place the testis back inside the body wall.

5. Repeat on the right testis.

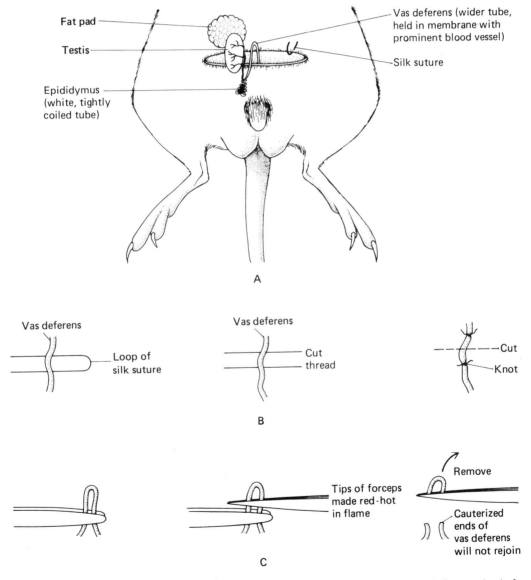

Figure 50 Two methods for vasectomizing a male mouse. In *A* the vas deferens is tied and then cut. In *B* the vas deferens is cauterized.

6. Sew up the body wall with two to three stitches. Then sew up the skin or clip the skin together with autoclips.

An alternative procedure for vasectomies is described in Rafferty (1970). Another very reliable technique is to proceed as described above to step 4. Then pull out the testis and associated organs with blunt forceps. Hold the vas deferens in a loop with one pair of forceps and cauterize with the red-hot tips of a second pair of forceps, as shown in Figure 50C.

WEEK SIX :   SUPEROVULATION AND BLASTOCYST TRANSFER TO
              PSEUDOPREGNANT FEMALES

Embryos from the one-cell stage through to the blastocyst stage (0.5–3.5 days p.c.) can be transferred into the reproductive tract of a pseudopregnant recipient to complete their development. Embryos from the one-cell through the morula stage (0.5–2.5 days p.c.) are transferred into the ampullae of 0.5-day p.c. pseudopregnant recipients, whereas 3.5-day blastocysts are transferred into the uterine horns of a 2.5-day p.c. (or at the very latest a 3.5 day p.c.) pseudopregnant recipient. The former procedure is called an oviduct transfer and can only be performed with embryos enclosed within a zona pellucida. The latter procedure is referred to as a uterine transfer and for this type of transfer the embryos need not have a zona pellucida. The reason for transferring embryos into a female at an earlier age of pseudopregnancy is to give the embryo time to "catch up" in its development before being exposed to conditions favorable for implantation. In general, for both procedures, 50–75% of unmanipulated embryos will develop to term. Ideally, enough embryos should be transferred to give a litter size of five to seven. If there are only one or two embryos in the uterus they may grow too big to be born without being damaged. Also, mothers may not take care of small litters. If the litters are too large (more than 10) then a few may grow up small, with a risk of being sterile. Provided that they have been given embryos injected with the same DNA, two foster mothers can be placed in the same cage after the transfer operation. They will subsequently help each other bring up a joint litter.

Oviduct Transfer

EQUIPMENT

Avertin (anesthetic)

Animal balance

1-cc syringe

26-gauge, 1/2-in. needle

Animal clippers (optional)

70% ethanol

Tissues (several rolled into small swabs are useful for soaking up
any blood)

Dissection scissors

Two #5 watchmaker's forceps (clean and sharp)

One blunt fine forceps

Surgical silk (size 5-0)

Curved surgical needle (e.g., size 10, triangular, pointed)

Serafine (1.5 in. or smaller)

Two steromicroscopes ideally (one for the operation, one for loading the embryo transfer pipet)

Fiber optic illuminator

Transfer pipet and mouth pipet

Alcohol burner

Wound clips and clip applicator

Lid of 9-cm plastic petri dish or glass plate

The recipients should be females, at least 6 weeks of age and >20 g in weight, mated to vasectomized males the evening before the transfer. Therefore, the females will be approximately 0.5 day p.c. pregnant at the time of the transfer. F_1 hybrid (e.g., C57BL/6×CBA or C57BL/6×DBA) or outbred females, such as Swiss albino mice, make good foster mothers. It is a good idea to have excess pseudo-pregnant females available in case the dissection of the oviducts is unsuccessful.

PROCEDURE

It is best to practice this procedure on a cadaver, and to inject a dye solution rather than eggs to convince yourself that you can find the opening to the oviduct. It is useful to remember that the position of the infundibulum is relatively invariant from mouse to mouse; with a little practice the technique will become routine and will cause minimal distress to the mouse.

1. Anesthesize the recipient. Weigh the recipient mouse and inject it intraperitoneally with 0.015–0.017 ml of 2.5% Avertin (Section H) per gram of body weight. (The proper dose of Avertin may vary with different preparations, and should be redetermined each time a new stock is prepared.) Place the mouse on a lid of a petri dish (or similar item) so that it can be lifted onto the microscope stage easily.

2. Shave the lower back of the recipient mouse (optional).

3. Load a transfer pipet with embryos. Since they will be outside the incubator for several minutes, transfer any embryos in M16 into M2 before loading. Transfer pipets are pulled in advance from BDH hard glass capillary tubes. The narrow part of the pipet should be 2–3 cm in length and 120–180 μm in diameter, i.e., just larger than one embryo and smaller than two. Also, the tip should be flush, and should be flame-polished, in order to minimize damage to the oviduct. The pipet is first filled with light paraffin oil to just past the shoulder (Fig. 51). The viscosity of the oil allows one to pick up or blow out the embryos with greater control. As an alternative to using oil, the other end of the pipet can be melted down to a narrow opening (see Implantating Tissues under the Kidney Capsule).

    A small amount of air is taken up, then medium M2, and then a second air bubble. Next the embryos are drawn up in a minimal volume of medium, then a third air bubble is taken up, followed by a short column of medium (see Fig. 51). Store the transfer pipet (still in the mouth-pipeting device) by pressing it into a piece of plasticene stuck to the stereomicroscope, and leave it there until you are ready to place the embryos in the oviduct. BE CAREFUL NOT TO DISTURB THE PIPET.

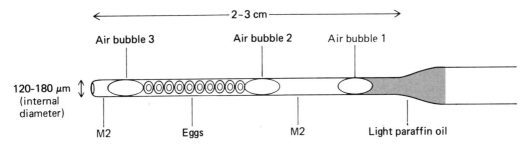

Figure 51 Embryo transfer pipet showing arrangement of air bubbles, medium, and embryos. Monitoring the position of the air bubbles enables the operator to be sure that all of the embryos have been injected.

4. Transfer the embryos.

   a. Sterilize all instruments by dipping them in 100% ethanol and flaming them with the alcohol burner. After wiping the mouse's back with 70% ethanol, make a small transverse incision (less than 1 cm) with the dissecting scissors, about 1 cm to the left of the spinal cord, at the level of the last rib (Fig. 52). Wipe the incision with 70% alchohol to remove any loose hairs.

   b. Slide the skin around until the incision is over the ovary (orange) or fat pad (white), both of which are visible through the body wall. Then pick up the body wall with the watchmaker's forceps and make a small incision just over the ovary. Stretch the incision with the scissors to stop any bleeding. With a surgical needle, thread a piece of silk suture through the body wall so that the body wall will be easy to locate later. With the blunt forceps, pick up the fat pad and pull out the left ovary, oviduct, and uterus, which will be attached to the fat pad. Clip the serafine onto the fat pad and lay it down over the middle of the back, so that the oviduct and ovary remain outside the body wall.

   c. Gently pick up the mouse and place it with head to the left on the stage of the stereomicroscope. This procedure is easier if the mouse is laid out initially on the lid of a petri dish or on a glass plate.

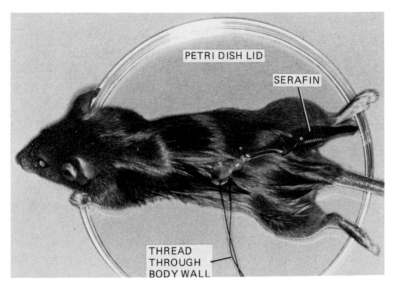

Figure 52 Mouse prepared for oviduct transfer. (A) The anesthetized mouse is placed on a petri dish lid for ease of handling. Ovary, oviduct, and proximal end of the uterus are held outside the body by means of a serafine clip attached to the fat pad above the ovary. The incision in the body wall is located with a suture thread. (B) The ovary/oviduct/uterus are held outside the body cavity by a serafine attached to the fat pad above the ovary. The orientation shown here gives easy access to the oviduct for transfer of embryos by the technique described in the text. After returning the uterus, etc., to the body cavity, the incision is located and closed by the thread through the body wall. In this demonstration the hair around the incision has not been shaved; this is an optional step that may be helpful when carrying out the procedure for the first few times.

d. Under the stereomicroscope locate the opening (infundibulum) to the oviduct and the swollen ampulla underneath the bursa (a transparent membrane over the oviduct and ovary). Arrange the mouse, oviduct, etc., so that the pipet can enter easily (Fig. 53). It is most convenient to have the head to the left and the ovary, etc., pulled to the rear with the serafine. With two watchmaker's forceps, tear a hole in the bursa over the infundibulum. Be careful not to tear through any large blood vessels.

e. Pick up an edge of the infundibulum or the bursa near the infundibulum with fine forceps and then insert the pipet down the opening to the ampulla. Blow until both air bubbles 2 and 3 have entered the ampulla.

f. Unclip the serafine and remove the mouse from the stereomicroscope. With the blunt forceps, pick up the fat pad and push the uterus, oviduct, and ovary back inside the body wall. Sew up the body wall with one or two stitches (optional) and close the skin up with wound clips.

5. Repeat steps 3 and 4 to transfer additional embryos to the right oviduct, if desired.

6. At the end of the operation, return the mouse to its cage and leave undisturbed in a warm, quiet place. It should recover from the anesthetic in about 20–30 minutes.

Rafferty (1970) and Dickman (1971) describe slightly different procedures for oviduct transfer.

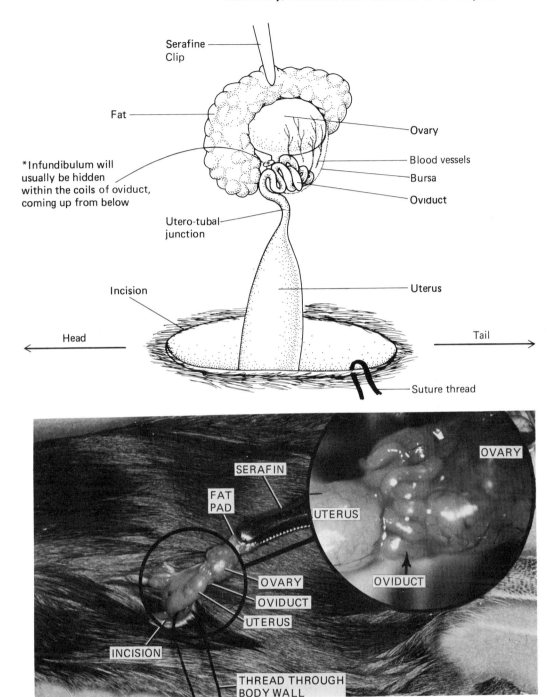

Figure 53 Detail for oviduct transfer. The oviduct is surrounded by a thin transparent bursa or membrane containing blood vessels. Make a small incision in the membrane (avoiding blood vessels), locate the end of the oviduct (infundibulum), and hold the edge with fine forceps while inserting the tip of the transfer pipet.

Uterine transfer

EQUIPMENT

> Avertin (anesthetic)
> Animal balance
> 1-cc syringe
> 26-gauge 1/2-in. needle
> Animal clippers (optional)
> 70% ethanol
> Tissues
> Dissection scissors
> Two #5 watchmaker's forceps
> One fine pair of blunt forceps
> Surgical silk (size 5-0)
> Surgical needle
> Serafine 4 cm or smaller (optional)
> Two stereomicroscopes (one for the operation; one for loading the
>   embryo transfer pipet)
> Fiber optic illuminator
> Mouth pipet
> Transfer pipet
> Alcohol burner
> Wound clips and applicator
> One 25-gauge 1.3 in. needle or sewing needle

Recipients are F_1 hybrid or outbred females, at least 6 weeks old, mated to vasectomized males. F_1 hybrid females (e.g., C57BL/6×CBA or C57BL/6×DBA) or outbred females such as Swiss albino mice have been found to make good recipients and foster mothers. The pseudopregnant females may be used as recipients at 2.5 or 3.5 days p.c. However, a 2.5-day pseudopregnant female is preferable since it allows the manipulated embryo plenty of time to catch up developmentally. Again, it is best to practice the procedure first on a cadaver, so that you will be more confident when you use an anesthetized mouse for the first time.

PROCEDURE

1. Anesthesize the recipient. Weigh the recipient mouse and inject it intraperitoneally with 0.015–0.017 ml of 2.5% Avertin per gram of body weight.

2. Shave the lower back of the mouse. This is optional and can be omitted when expertise is achieved.

3. Expose the uterus.

    a. Sterilize all instruments by dipping in 100% ethanol and flaming with the alcohol burner.

    b. After wiping the mouse's back with 70% ethanol, make a single small longitudinal incision (less than 1 cm) with the dissecting scissors in the midline at the level of the last rib (Fig. 54). Wipe the incision with 70% alcohol to remove any loose hairs.

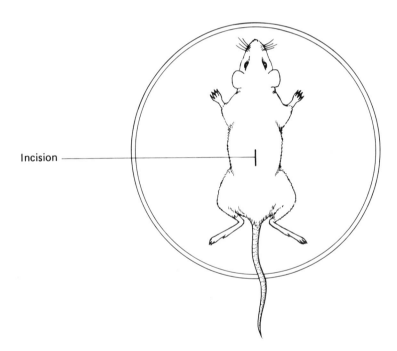

Figure 54   Position of incision for uterine transfers.

c. Slide the skin to the left or right until the incision is over the ovary or fat pad, both of which are visible through the body wall. Then pick up the body wall with the watchmaker's forceps and make a small incision just over the ovary. Stretch the incision with scissors to stop any bleeding. Insert a piece of surgical silk through the body wall so it will be easy to locate later.

d. With the blunt forceps, grab the fat pad and pull out the ovary, oviduct, and uterus, which will be attached to the fat pad (Fig. 55). If required, clip a serafine onto the fat pad and lay it down over the middle of the back, so that the oviduct and ovary remain outside the body wall.

e. Gently pick up the mouse and place it on the stage of the stereomicroscope.

4. Load a transfer pipet with about seven blastocysts. This will give about five embryos coming to term (75% success rate).

a. Transfer pipets are prepared in advance from a Pasteur pipet or a BDH hard glass capillary by pulling it out to about 200 μm in diameter, and then fire-polishing the end. A bend is placed in the pipet about 1 cm from the end. This allows one to judge how far the pipet has been inserted into the uterus.

b. The pipet is first filled with light paraffin oil up to the shoulder. A small amount of air is taken up (air bubble 1), then DMEM or similar medium, and then air bubble 2. Next the blastocysts are picked up in a minimal volume of medium (filling about 0.5 cm of the pipet). No third air bubble is taken up in the pipet (air bubbles in the uterus may interfere with implantation).

c. Store the pipet by pressing it into a piece of plasticene stuck to the stereo-microscope, until it is time to place the embryos in the uterus. BE CAREFUL NOT TO DISTURB THE PIPET.

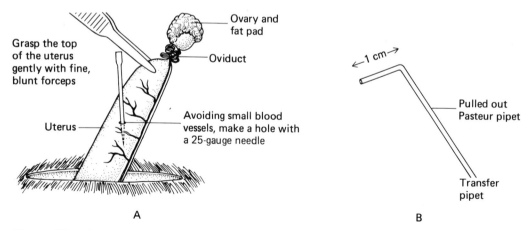

A   B

Figure 55 Diagram showing technique of uterine transfer (*A*). A 26-gauge or sewing needle is used to make a hole all the way into the lumen of the uterus. After removing the needle the transfer pipet (*B*) is inserted into the hole and the blastocysts injected.

5. Transfer the embryos.

   a. Hold the top of the uterus gently with the fine blunt forceps and with a 26-gauge or sewing needle make a hole a few millimeters down the uterus. Avoid small blood vessels in the uterine wall. Make sure the needle has entered the lumen and has not become lodged in the wall of the uterus. To test whether the needle has entered the lumen, pull it out slightly. If it slides easily, the needle has penetrated the lumen. Do not move the needle around too much or the wall of the uterus may be lacerated.

   b. Keeping your eye on the hole made by the needle, pull out the needle and insert about 5 mm of the transfer pipet. Blow gently until the air bubble closest to the embryos (air bubble 2) is at the tip of the pipet and all of the blastocysts have been expelled.

   c. Unclip the serafine and remove the mouse from the microscope stage. With the blunt forceps, pick up the fat pad and place the uterus, oviduct, and ovary back inside the body wall. Sew up the body wall with one or two stitches (optional) and close the skin with wound clips.

6. Repeat steps 3c, 4, and 5 on the right side, if desired.

Alternative methods for uterine transfer are described in Mintz (1967) and Rafferty (1970).

CAESARIAN SECTION AND FOSTERING

EQUIPMENT

> Scissors
> Blunt forceps
> Anglepoise lamp
> Tissues

Caesarean section is necessary if the recipient female has not given birth to her pups by the time normal for the particular mouse strain. This often occurs when only one or two embryos are present and they grow too big to be born.

PROCEDURE

1. If a suitable foster mother is available, kill the pregnant female by cervical dislocation and quickly open the abdomen and cut out the uterus. Dissect the pups from the visceral yolk sac and amnion and wipe away fluid and secretion with tissues, particularly from mouth and nostrils. Cut the umbilical cord and gently pinch the body with blunt forceps to stimulate breathing. Place on damp tissues under a lamp to keep warm.

2. The next step is to persuade the foster mother to accept the babies as her own and to take care of them. An ideal foster mother is a female who has successfully reared one or more litters of her own and has given birth the same day or the previous day. (If a good supply of breeding females is not routinely available, it is a good idea to set up normal matings at the same time that the pseudopregnant females are mated.) Use foster mothers of a different coat color or, alternatively, clip off one of the toes of all the natural babies, so that the fostered babies can be distinguished later. Remove the mother from the cage and mix the foster babies with the natural litter and bedding. Induce the mother to urinate on the whole lot and again mix them together before returning the mother to the cage. Some workers say that light etherization of the foster mother makes her "forget" her babies and therefore accept the new ones more readily (C.F. Graham, pers. comm.). Another useful tip is to minimize the temperature difference between the two sets of babies; if the caesarian baby is too cold it may be recognized as "dead" by the mother and rejected. If necessary, some of the natural babies can be removed over the next few days to leave a litter size of about 7 or 8.

IMPLANTING TISSUES UNDER THE KIDNEY CAPSULE

This technique* can be used to test whether adult or embryonic tissues will grow in vivo without rejection, and to obtain teratocarcinomas from early embryos.

EQUIPMENT

Avertin anesthetic
Desmarres chalazion forceps (see Section H, Supplies)
Stereomicroscope
Watchmaker's forceps, wound clips, etc.
"Breaking pipet" if a small piece of tissue is to be transferred
Sterile saline or phosphate-buffered saline
Isotonic

*Information provided by Dr. Davor Solter, Wistar Institute, Philadelphia, Pennsylvania 19104.

PROCEDURE

1. Anesthetize the mouse and make a ∿1-cm cut in the skin, as shown in Figure 54. Slide the incision to one side and cut the body wall just above the level of the ovary. Pull out the kidney by its fat pad, using blunt forceps. Immobilize the kidney in the chalazion forceps. Allow the surface of the kidney to dry for a few minutes. (This enables the capsule to be picked up with fine watchmaker's forceps.)

2. Make a small tear in the capsule membrane with watchmaker's forceps and then moisten the capsule with saline. Make a pocket underneath the capsule with cleaned watchmaker's forceps (moistened) and then insert the tissue to be grown with the forceps or with a "breaking" pipet. The function of this pipet is to prevent backflow of blood into the pipet which may result in clogging of the opening.

   a. To make a breaking pipet, pull out a glass capillary on a microburner to give a very fine tip. Insert the end of the capillary with the fine tip into a larger piece of glass tubing using, for example, a Drummond micropipet tip or sealing wax (see Fig. 56). Pull out the other end of the capillary to the desired dimensions. Basically, the resistance to the flow of liquid will be inversely proportional to the diameter of the fine end of the capillary.

3. Push the tissue as far away from the tear as possible. Release the kidney from the forceps and replace in the body cavity with blunt forceps. Close the body wall and skin with sutures and wound clips.

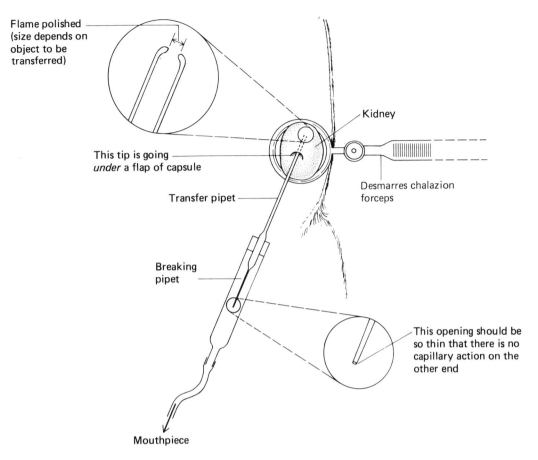

Flame polished
(size depends on
object to be
transferred)

Kidney

This tip is going
under a flap of capsule

Desmarres chalazion
forceps

Transfer pipet

Breaking
pipet

This opening should be
so thin that there is no
capillary action on the
other end

Mouthpiece

Figure 56 Technique for implanting a small piece of tissue under the kidney capsule using a breaking pipet.

Section D

INTRODUCTION OF NEW GENETIC INFORMATION INTO THE DEVELOPING MOUSE EMBRYO

During the past decade, a number of techniques for manipulating the mammalian embryo have been developed that allow the stable introduction of new genetic material into the germ line. These techniques have revolutionized mouse genetics by making it possible to transfer any cloned gene, normal or mutant, into the germ line and subsequently to study the expression of the introduced gene during development. In addition, these techniques should allow the genetic manipulation of agriculturally important animals.

INTRODUCTION

Why Transfer Genes into the Mouse?

First, it has become possible to use gene transfer to study the molecular basis of tissue-specific and stage-specific gene expression during mouse development. Most cloned genes that have been introduced into the mouse germ line have shown appropriate tissue-specific and stage-specific patterns of expression (e.g., Grosschedl et al. 1984; Storb et al. 1984; Swift et al. 1984; Chada et al. 1985; Hanahan 1985; Krumlauf et al. 1985; Magram et al. 1985; Overbeek et al. 1985; Shani 1985; Townes et al. 1985; for review, see Palmiter and Brinster 1985), despite their integration into apparently random sites in the host genome. Therefore, gene transfer into the mouse embryo, combined with in vitro mutagenesis, can serve as an experimental assay for the *cis*-acting DNA sequences that dictate specific patterns of transcription in the developing animal.

Second, the ability to introduce and express cloned genes allows one to investigate the phenotypic effects of altered gene expression. One can cause the overproduction of a normal gene product, or its synthesis in an unusual cell type, by attaching the mRNA encoding portion of the gene to novel regulatory sequences (Palmiter et al. 1982); conversely, it may ultimately be possible to inhibit the expression of an endogenous gene in a developing mouse, by introducing a gene encoding antisense RNA (Izant and Weintraub 1985). These approaches can be used either with a gene whose normal function is already known (e.g., growth hormone; Palmiter et al. 1982), or with newly identified genes whose function is being investigated. In addition, novel genes not normally expressed in the mouse (e.g., SV40 T antigen; Brinster et al. 1984; Hanahan 1985), or mutant forms of normal mouse genes, can be introduced. These approaches provide new experimental strategies for answering basic questions in many areas of mammalian biology, and may also allow the production of animal models of human diseases.

Third, it is possible to use foreign DNA as an insertional mutagen which, unlike conventional mutagens, "tags" the mutated gene and allows it to be molecularly cloned and characterized. Foreign genes, or retroviruses, introduced into the mouse germ line have caused a variety of mutations, most often recessive (Jaenisch et al. 1983; Schnieke et al. 1983; Wagner et al. 1983), with phenotypes ranging from developmental arrest at the late blastocyst stage (Wagner et al. 1983; Mark et al. 1985) to limb deformities (Woychik et al. 1985). The frequency of such mutations appears to be in the range of 5–15%. Although it is not feasible to screen for insertional mutations that cause a particular phenotype, the large number of mutations that are generated in transgenic mice (many as by-products of other experiments), combined with the ability to clone the affected loci, should provide a unique opportunity to identify new genes whose expression is essential for normal mammalian development.

How to Transfer Genes into the Mouse

The method that has been most extensively, and most successfully, employed is the microinjection of DNA directly into the pronuclei of fertilized mouse eggs (Gordon et al. 1980). This leads to the stable chromosomal integration of the foreign DNA in 10–40% of the surviving embryos (Brinster et al. 1981b; Costantini and Lacy 1981; Gordon and Ruddle 1981; E. Wagner et al. 1981; T. Wagner et al. 1981; Lacy et al. 1983). In most cases, integration appears to occur at the one-cell stage, since the foreign DNA is present in every cell of the "transgenic" animal, including all the primordial germ cells. In about 20% of transgenic mice, the foreign DNA apparently integrates at a later stage, resulting in animals that are mosaic for the presence of the foreign DNA (e.g., Costantini and Lacy 1981). The number of copies of the foreign DNA sequence that are retained (per cell) can range from one to several hundred, and bears little relation to the number of DNA molecules injected into the egg. When multiple copies are present, they are normally found at a single chromosomal locus; however, occasionally there may be separate integration sites on two different chromosomes (Lacy et al. 1983). The copies of the foreign DNA at each integration site are arranged primarily in a head-to-tail array (Brinster et al. 1981b; Costantini and Lacy 1981), although some integration events may be more complicated (Wagner et al. 1983). The sites of integration within the host genome appear to be diverse (Lacy et al. 1983).

An alternative method for introducing genes into the mouse germ line is the infection of embryos with retroviral vectors. Pre- or postimplantation embryos can be infected by either wild-type or recombinant retroviruses, leading to the stable integration of viral genomes into the host chromosomes (Jaenisch 1976; Jaenisch et al. 1981; Stuhlmann et al. 1984; Jahner et al. 1985; van der Putten et al. 1985). Retroviral integration occurs through a precise mechanism, leading to the insertion of single copies of a provirus; the tandem arrays obtained when DNA is injected are never observed. When preimplantation embryos (four- to eight-cell) are infected, independent proviral integrations can occur in one or more cells, leading to an animal that is mosaic for each integration; in addition, multiple integration events can occur in any one cell. The frequency of obtaining transgenic animals can be as high as that obtained by DNA microinjection, although it varies greatly depending on the titer of virus used (van der Putten et al. 1985). Germ line transmission rates are generally low, due to extensive mosaicism (Jaenisch et al. 1981). When midgestation embryos are infected, germ line transformation is extremely rare (Jaenisch et al. 1981; Stuhlmann et al. 1984).

One advantage of DNA microinjection is that, given the proper equipment and expertise, it is a faster and simpler procedure. Any cloned DNA sequence, regardless of size (up to at least 50 kb; Costantini and Lacy 1981), can be introduced directly into the mouse genome by microinjection into the zygote. In contrast, there is a limitation on the size of a DNA fragment that can be inserted into a retroviral vector (about 10 kb), and the preparation of a DNA fragment for introduction into the embryo by viral infection is more involved. The cloned gene to be transferred must first be inserted into a viral vector, and then transfected into a cell line for the production of recombinant virions. This is not a serious limitation if one or a few genes are to be transferred, but it could be cumbersome if, for example, expression in transgenic mice is to be used as a routine assay for regulatory DNA sequences. Another advantage of DNA microinjection is that it usually results in nonmosaic transgenic mice that carry the foreign DNA sequences in every cell. This facilitates the analysis of gene expression in the "founder" transgenic mouse, and also allows the relatively rapid development of a transgenic "line." In contrast, transgenic mice produced by retroviral infection are extensively mosaic, and several generations of breeding may be required before a homogeneous transgenic line can be obtained.

The use of viral vectors, on the other hand, eliminates the need for expensive microinjection equipment. Perhaps more importantly, it allows the generation of transgenic mice carrying a single, intact copy of a foreign gene. Large tandem arrays represent an abnormal organization for most genes and this arrangement, and/or a high copy number, might preclude their normal regulation. While genes introduced in such an arrangement are usually expressed tissue specifically, their levels of expression are extremely variable and rarely approach the appropriate level, if normalized for gene copy number. It is possible that genes introduced into mice via retroviral vectors will be expressed more normally, as are genes introduced into *Drosophila* using P-element vectors (Rubin and Spradling 1982), but this remains to be demonstrated. It should be noted that while transcription from retroviral promoters is permanently repressed following integration into early mouse embryos (Jahner et al. 1982, 1985), or embryonal carcinoma cells (Stewart et al. 1982; Gautsch and Wilson 1983; Niwa et al. 1983), transcription from internal promoters in recombinant retroviruses appears not be to similarly inhibited (Rubenstein et al. 1984; Wagner et al. 1985). At this time, the use of retroviral vectors to introduce and express foreign genes in developing mice remains to be perfected, while the use of DNA microinjection has been widely applied with considerable success.

For purposes of insertional mutagenesis, the use of recombinant retroviruses may also be advantageous (Jaenisch et al. 1983). It is relatively easy to clone a gene that has been inactivated by a retroviral insertion, particularly if the virus is tagged with a suppressor gene to allow selection in *Escherichia coli* (King et al. 1985; Lobel et al. 1985). In contrast, genes inactivated by the integration of microinjected DNA may be more difficult to recover, particularly if integration involves extensive rearrangement (Wagner et al. 1983); however, at least one such gene has been successfully recovered (Woychik et al. 1985). On the other hand, only one mutation in the mouse has so far been generated by experimental infection with a retrovirus (at midgestation; Jaenisch et al. 1983), while mutations caused by DNA insertion occur relatively frequently.

An entirely different method of transferring new genetic information into the mouse embryo involves the introduction of embryonic stem cells into the preimplantation embryo. Teratocarcinoma-derived embryonal carcinoma (EC) cells can colonize the embryo and participate in normal development when either injected into the blastocyst (Brinster 1974; Mintz and Illmensee 1975; Papaioannou et al. 1975) or aggregated with eight-cell embryos (Stewart 1982; Fujii and Martin 1983). However they very rarely contribute to the germ line (Mintz and Illmensee 1975; Cronmiller and Mintz 1978; Papaioannou et al. 1978; Stewart and Mintz 1982). More recently, embryonic stem cell lines have been derived in vitro from normal blastocysts (EK cells, Evans and Kaufman 1981; ES cells, Martin 1981). These cells have been shown to colonize the germ line regularly, as well as the somatic tissues, when introduced into the embryo (Bradley et al. 1984). These techniques, combined with the ability to introduce cloned genes into EK cells, should allow an alternate route for gene transfer into the germ line. However, this approach has been unsuccessful so far; although cloned genes have been introduced into EK cells by several techniques, such genetically altered EK cells have so far failed to contribute to the germ line (M. Evans; E. Wagner; both pers. comm.), perhaps due to chromosomal abnormalitites acquired during selection in vitro. If this approach were to be successful, a possible advantage over direct introduction of a gene by microinjection or viral infection might be the ability to preselect cell lines with desirable properties (i.e., high expression of the foreign gene) before introducing them into the embryo.

Perhaps a more important application of this approach will be the ability to introduce preselected mutations in specific genes into the mouse germ line (Dewey et al. 1977; Mintz 1978; Watanabe et al. 1978). It is extremely laborious and expensive to generate and identify mutations in specific genes in the mouse, even using chemical mutagens. Insertional mutagenesis of a specific gene of choice in the mouse is not feasible. In contrast, cells grown in culture can be readily subjected to mutagenesis and screened for specific mutations. Mutant EC cells can be introduced into the developing mouse, and Mintz and her colleagues have used this approach to produce mice containing hypoxanthine-guanine phosphoribosyl transferase (HPRT)-deficient cells (Dewey et al. 1977); unfortunately, the mutation was not passed through the germ line. Possibly similar experiments with EK cell lines may be more successful.

New genetic material may also be introduced into the mouse embryo by nuclear transplantation, although this approach does not allow the introduction of specific cloned genes or mutations. Rather, this technique has been used to examine the totipotency of early embryonic nuclei, the nuclear versus cytoplasmic transmission of genetic defects, and the equivalence of maternal and paternal genomes in early development (Hoppe and Illmensee 1977, 1981; Illmensee and Hoppe 1981; McGrath and Solter 1983a,b, 1984a,b; Surani and Barton 1983; Surani et al. 1984).

MICROINJECTION OF DNA INTO PRONUCLEI

EQUIPMENT

Light paraffin oil or Fluorinert Electronic Liquid FC77 (3M Company)

5-cm-long, 26-gauge needles for loading holding pipet (e.g., Hamilton, catalog number KF726/90126)

Glass tubing for injection pipet, e.g., capillaries with an internal glass filament (from W-P Instruments Inc., New Haven Connecticut, catalog number TW100F; or Clark Electromedical Instruments, Pangbourne, Reading RG8 7HU, U.K., catalog number GC 100TF-15)

Glass tubing for making holding pipet, 1-mm o.d. (Leitz glass capillaries, E. Leitz, catalog number 520119)

Glass depression slides (e.g., Fisher 12-560A) or 90-mm plastic tissue culture dishes

50-cc glass syringe with ground glass plunger

Inverted microscope with a 40× long-working-distance objective, low-power 4× or 10× objective, and a long-working-distance condenser (see The Microinjection Setup)

Right-hand and left-hand Leitz micromanipulators - s terling

Baseplate (see Appendix 1)

Two Leitz single instrument holders

Leitz instrument tubes (several)

Isolation table (if necessary) (e.g., Micro-G table, Technical Manufacturing Corp., Woburn, Massachusetts, or Moore and Wright, Sheffield, England)

Micrometer syringe (e.g., Agla micrometer syringe, catalog number MS01 Wellcome Research Labs, Beckenham, BR3 3BS U.K.)

Mechanical pipet puller (e.g., Model 700/D or 720 David Kopf Instruments, Tujunga, California)

Microforge (e.g., DeFonbrune, obtainable through Microinstruments (Oxford) Ltd., Oxford OX1 2HP or Kramer Scientific Corp., Yonkers, New York 10701); the microforge should be fitted with a thick filament 0.22 mm or 32- or 36-gauge platinum wire, 10×, 20× lenses, 10× focusing eyepiece with graticule

Tygon tubing (e.g., 3/32 in. i.d., 5/32 in. o.d.)

Two clamp stands

Diamond pencils

Factors Affecting the Efficiency of Gene Transfer

Experiments to date suggest that virtually any DNA molecule can be efficiently inserted into the mouse genome by microinjection into the zygote. Although the frequencies of integration obtained with different DNA fragments may vary by three- to fourfold under seemingly identical conditions, it is not clear whether this is an effect of DNA sequence or of some other variable such as the purity of the DNA preparation. The length of a DNA sequence that can be introduced is currently limited only by molecular cloning technology: 50-kb λ clones (Costantini and Lacy 1981) or cosmid clones (Y.W. Kan, pers. comm.) can be introduced in their entirety, with high efficiency.

   Several properties of the injected DNA may affect its frequency of integration and/or the structure of the integrated DNA sequences. The properties that can be most easily controlled are discussed below, and include:

1. The linear or circular form of the DNA.

2. The ends of a linear molecule: blunt, single-stranded cohesive, or single-stranded noncohesive ends.

3. The DNA concentration.

4. The buffer in which the DNA is dissolved.

5. The purity of the DNA.

Linear or Circular Form

While transgenic mice have been produced by microinjection of either linear or supercoiled DNA molecules, linear molecules appear to integrate in a larger fraction of injection zygotes. In the only controlled study that has been published, Brinster et al. (1985) reported that linear DNA molecules gave a fivefold higher integration frequency than supercoiled molecules (25% versus 5%). Microinjection of either linear or circular forms most frequently results in the integration of multiple copies of the injected DNA in one or more head-to-tail arrays. This is likely to occur by rapid circularization of linear molecules after injection into the pronucleus, followed by homologous recombination between several circular molecules and integration into the genome (Brinster et al. 1981b, 1982).

Blunt, Single-stranded Cohesive, or Single-stranded Noncohesive Ends

The structure of the ends of a linear molecule appears to have little effect on either the frequency of integration or the structure of the integrated molecules. Brinster et. al. (1985) observed a severalfold lower frequency for blunt ends (filled in with the Klenow fragment of DNA polymerase) than for single-stranded ends (generated either with the same or different restriction enzyme): 8% (5/59) for blunt ends, 24% (5/21) for similar ends, and 31% (13/42) for dissimilar ends. However, in another study, a frequency of 39% (5/13) was obtained with blunt ends (Grosschedl et al. 1984). Regardless of the type of ends a linear DNA fragment carries, the injected molecules will most frequently integrate as a head-to-tail array.

DNA Concentration

It is difficult, if not impossible, to control accurately the volume of DNA solution introduced into the pronucleus; most investigators estimate that 1–2 pl is introduced, but the fraction of DNA that remains in the nucleus is unknown. There-fore, the only variable that can be controlled is the concentration of the DNA solution that is injected, and this appears to be an important variable. Brinster et al. (1985) found that optional DNA integration efficiency (20–40%) was achieved with concentrations of linear DNA of 1 μg/ml or higher. However at high DNA concentrations (>10 μg/ml) embryo survival decreased significantly. Therefore, optimal numbers of transgenic mice were produced when DNA was injected at a concentration of 1–2 μg/ml, corresponding to 200–400 molecules/pl of 5-kb DNA fragment. Similarly, linear DNA fragments with sizes ranging from 5 kb to 8 kb have been found to integrate with about a 5% frequency when injected at 0.1–0.2 μg/ml, and a frequency of 20–40% when injected at concentrations of 1–3 μg/ml (F. Costantini, unpubl.).

Unfortunately, there appears to be little if any correlation between the concen-tration of DNA injected and the number of copies that integrate. By injecting DNA at a low concentration (20 molecules/pl), we were able, in one experiment, to gen-erate transgenic mice most of which (5 out of 7) carried single copies of the in-jected DNA molecule (Chada et al. 1985). However, in most experiments the injec-tion of DNA at low concentrations most frequently yields mice containing multiple copies of the injected DNA, and only reduces the frequency of transgenic mice (F. Costantini, unpubl.).

Buffer in Which DNA Is Dissolved

Injection buffer typically contains 5–10 mM Tris (pH 7.4) and 0.1–0.25 mM EDTA. Addition of 1 mM EDTA or 1 mM MgCl$_2$ reduced embryo survival by 30–50%, while addition of 5 mM EDTA or 3 mM MgCl$_2$ was extremely toxic; elimination of EDTA led to reduced survival and reduced integration efficiency (Brinster et al. 1985). The effect of buffer composition on other parameters, such as DNA copy number, has not been investigated.

Purity of DNA

DNA samples for microinjection should, of course, be free of contaminants that might harm the eggs, such as traces of phenol, ethanol, or enzymes. In addition, it is essential to get rid of any particulate matter that could clog the injection needles. For this reason, all solutions added to the DNA sample should be filtered through a 0.2-μm filter, and all tubes and pipets should be rinsed with filtered water prior to use. A suggested DNA purification procedure is described in the next section.

While prokaryotic cloning vector sequences have no apparent effect on the integration frequency of microinjected genes, it is important to note that they can severely inhibit the expression of eukaryotic genes introduced into the mouse germ line. This effect was initially reported by Chada et al. (1985), who observed that a cloned β-globin gene was expressed tissue specifically in transgenic mice only when the plasmid sequences were removed before injecting the gene. Similarly, Townes et al. (1985) observed a 100- to 1000-fold increase in the level of expression of the human β-globin gene when plasmid sequences were removed prior to microinjection. Many genes other than globin have been expressed in an appropriate tissue-specific pattern in transgenic mice, despite the presence of prokaryotic DNA sequences in the introduced DNA fragment. Nevertheless, in several of these cases (including the α-fetoprotein gene, S. Tilghman, pers. comm.; a metallothionein–growth hormone gene and an elastase–growth hormone fusion gene, R. Palmiter, pers. comm.), much higher levels of transcription and more reproducible patterns of expression were observed when plasmid sequences were removed prior to microinjection. It is not known whether a specific sequence is responsible for this inhibition, or whether it is a general property of prokaryotic DNA. In any case, it is advisable to remove all vector sequences from a cloned gene before introducing it into the mouse germ line, if optimal expression of the gene is desired.

Preparing DNA Samples for Microinjection

Two alternative methods for purification of DNA for microinjection are provided: (1) using CsCl centrifugation and (2) using glass powder.

Purification of DNA for Microinjection by CsCl Centrifugation

1. Prepare the DNA fragment to be injected. This might involve linearizing a plasmid with a restriction endonuclease, or purifying a DNA fragment by electrophoresis on an agarose gel and electroelution.

2. Extract the DNA several times with phenol to remove enzymes, agarose contaminants, etc.

3. Extract with ether three times to remove residual phenol.

4. Precipitate with ethanol several times.

(All of these procedures are described in Maniatis et al. 1982).

   The DNA may now be pure enough to microinject following extensive dialysis against injection buffer (see above). Some DNA samples, particularly those eluted from agarose gels, may still contain particulate contaminants that will clog the injection needle. For an added purification step that will solve this problem, the DNA can be centrifuged to equilibrium in a CsCl density gradient as follows:

5. Dissolve the DNA (2–10 μg) in 2.4 ml of 10 mM Tris-HCl (pH 8), 1 mM EDTA. Add exactly 3.0 g of ultrapure CsCl (E.M. Reagents, catalog number 2039), let dissolve, and check that the density is 1.70 ± 0.01 g/ml.

6. Transfer to a clean 1.3×5-cm polyallomer ultracentrifuge tube, cover with a light paraffin oil, and centrifuge in a SW50.1 rotor (Beckman) at 20°C, 40,000 rpm for 48 hours.

7. Collect 0.2-ml fractions from the bottom of the tube. Assay eight fractions from

the middle of the tube by running 1–3 µl on an agarose minigel, and pool those fractions containing the DNA.

8. Dialyze at 4 °C against a large volume of injection buffer, changing the buffer several times over a 48-hour period.

9. Dilute the DNA in injection buffer to the desired concentration, and store frozen in 50-µl aliquots.

If clogging of the injection needles is a problem, some particulate matter may be removed by centrifuging the DNA samples in a microfuge (13,000g) or in a Beckman Airfuge, and discarding the bottom one-fourth of the tube contents. This step is usually not necessary for DNA samples that have been purified in CsCl as described above.

Isolation and Purification of DNA for Microinjection Using Glass Powder

In many cases it is desirable to utilize fragments of DNA purified from agarose gels for microinjection. Several methods for isolating or purifying these fragments, such as electroelution, phenol extraction, or column chromatography, often yield DNA that is still not pure enough for optimal results in microinjection. The following procedure* is one that is simple and may be used directly or in conjunction with other methods to provide DNA fragments that work with high efficiencies in microinjection experiments and for most other uses in molecular biology. It is based on a paper by Vogelstein and Gillespie (1979) modified by Bob Lyons. Double-stranded DNA binds to many substances in the presence of high salt. The concentration necessary differs for each salt, but 4 M is adequate for NaI. In addition, NaI is a chaotropic salt that can dissolve agarose. Glass powder is used to bind the DNA because it has a large surface area and high binding capacity. This method works with a high recovery efficiency even with small amounts of DNA (0.1 µg), and may be used for large-scale or preparative isolations. Fragments from 100 bp up to 40 kb have been isolated with nearly quantitative yields.

MATERIALS

1. Powdered flint glass 325 mesh. Available from most glass supply companies (example, Eagle Ceramics, Inc., 12267 Wilkins Avenue, Rockville, Maryland 20852).

2. Sodium iodide solution: 90.8 g NaI, 0.5 g $NaSO_3$. Dissolve in water to final volume of 100 ml. Filter through Nalgene 0.45-µm filter. Add back a few crystals of sodium sulfite. Note that the sodium sulfite does not dissolve completely. It stabilizes the solution against air oxidation. Store at 4 °C wrapped in aluminum foil to protect from light.

*Information supplied by Dr. Robb Krumlauf, National Institute for Medical Research, London, UK.

3. Ethanol wash solution: 50% ethanol, 0.1 M NaCl, 10 mM Tris (pH 7.5), 1 mM EDTA. Store at −20 °C.

4. Elution buffer: 10 mM Tris (pH 7.5), 1 mM EDTA.

5. 20× TAE gel buffer: 0.8 M Tris, 0.4 M NaAc, 0.02 M EDTA.

Preparation of the glass powder:

1. Resuspend 250 ml of glass in 500 ml of water. Stir well. Allow to settle for 1 hour.

2. Discard the settled (coarse) glass powder; centrifuge the supernatant to recover the fines.

3. Resuspend the glass powder in 200 ml of water, add 200 ml concentrated nitric acid. Bring to a boil in the hood.

4. Spin out the glass, resuspend in water to wash away the acid. Repeat this until the water is neutral. Optionally, you may wish to wash once with 1 M Tris (pH 7.5), but be sure to wash away the Tris afterwards.

5. Store in 1-ml aliquots at room temperature as a 50% slurry in water.

Note: The glass powder slurry can grow bacteria. You may wish to use sterile materials in its preparation and handle it appropriately afterwards.

Purification of DNA and/or recovery of DNA from agarose gels:

1. In the case of agarose gels, the gel can be run in normal agarose (e.g., BRL Ultra-Pure agarose), but MUST be run in TAE buffer, NOT a borate buffer. Excising the minimum-sized fragment of agarose speeds the procedure by allowing all steps to be performed in microfuge tubes and can improve recovery.

2. Weigh the gel to estimate volume and add 2 volumes of 6 M NaI. Incubate at 37 °C until the gel dissolves, about 15 minutes. Vortex periodically.

Note: In the case where the DNA does not need to be run on an agarose gel, or has already been isolated from the gel by an alternative method, dissolve the DNA in 10 mM Tris, 1 mM EDTA. Then add 2 volumes of 6 M NaI. It is not necessary to incubate the samples at 37 °C if there is no agarose to dissolve.

3. Vortex the glass suspension to reform the slurry and add 1 μl of glass powder suspension to the digested gel for every 2 μg of DNA present.

4. Chill on ice for 1 hour with occasional mixing.

5. Spin out the glass powder at as low a speed as possible (1–2 sec in microfuge). Discard supernatant.

6. Resuspend glass in NaI solution at half the original gel volume. Make sure any clumps of glass powder are dispersed. Spin out the glass powder (1–2 sec in microfuge). Repeat once more.

7. Resuspend the glass in ½ volume of ethanol wash solution and centrifuge to wash. Repeat once more.

8. Remove as much ethanol as possible with a fine microcapillary. Do NOT allow the glass pellet to dry. Immediately add elution buffer.

9. Elute the DNA by adding Tris/EDTA elution buffer and incubating at 37 °C for 15 minutes. Usually elute to give 10–100 ng/µl DNA; however, you have the option of eluting in whatever volume is convenient. (If the volume used is very small, i.e. 5–20 µl, a second elution of the powder is recommended with an equal volume of TE.)

10. Spin out the glass in microfuge (3 min) and discard. Save supernatant.

Note: At this stage the DNA is suitable for injection, nick-translation, etc. However, two alternative steps have been added to remove any remaining glass powder or traces of ethanol, and seem to aid in preventing blockage of injection pipets. Either: (1) Pass through a G-50 spun column equilibrated in TE. If the volume is less than about 200 µl, this can be done immediately. If the volume is large, ethanol precipitate first and resuspend in 100-200 µl of TE, or (2) filter through a small Millipore filter (catalog number SJHVL04NS, Type HY, 0.45 µm).

Finally, assay the eluate for DNA concentration and dilute appropriately. The concentration is conveniently assayed by comparing the ethidium bromide staining of a sample run on an agarose gel next to 100 ng of λ DNA cut with *Hind*III. Store the DNA at either 4° or −20 °C in 50-µl aliquots in clean Eppendorf tubes.

Choice of Mouse Strain

To obtain the largest number of fertilized eggs from the fewest females, many workers have used F_1 hybrid or outbred mice. Fertilized eggs from F_1 hybrid females may also survive micromanipulation and in vitro culture with a somewhat higher frequency than those from inbred strains. F_1 hybrids that have been used for microinjection experiments include C57BL/6×CBA, C57BL/6×DBA, C57BL/6×SJL, and C57BL/6×Balb/c. Fertilized eggs from inbred strains of mice (e.g., C57BL/6) have also been successfully used for DNA microinjection experiments. Although it is less convenient to use inbred mice, in terms of the numbers of eggs that can be easily obtained, embryo survival, and general reproductive performance, the subsequent benefits may outweigh the disadvantages if one intends to develop permanent transgenic mouse lines and analyze them genetically.

Timing of Injection

The pronuclei of fertilized mouse eggs can be most easily injected when they are at the maximum size before the nuclear membrane disappears. The pronuclei swell progressively during the one-cell stage, and they are in an optimal state for injection for a period of about 3–5 hours. The exact time of day when the eggs are optimal depends on the mouse strain being used, as well as the light-dark cycle on which the mice are maintained. For a light-dark cycle with lights off at 7 pm and on at 5 am, the eggs can be injected between about 2 pm and 7 pm. The best timing for injection should be determined empirically. Successful gene transfer does not appear to depend on injection at any specific time in the cell cycle. Fertilized eggs for microinjection can be obtained either by natural mating (Section B) or by induction of superovulation (Section C).

Making Holding Pipets

1. Hold a 10-cm piece of 1-mm o.d. Leitz capillary tubing with both hands and heat a small region at the center by rotating it in a microflame. When the glass becomes soft, pull the ends apart about 1 cm, then quickly remove it from the flame and pull the two ends sharply to draw out a thin region 5–10 cm long. The outside diameter of the drawn-out region should be about 80–120 μm, and definitely not more than 150 μm. The trick to obtaining very thin pipets is to draw out the capillary partially before removing it from the flame, and then to give a second hard pull out of the flame.

2. Next, score the glass very lightly with a diamond pencil about 2 cm from the shoulder of the pipet and bend it until it breaks at this point. Examine the pipet under the steromicroscope and make sure it has broken cleanly to give a perfectly flat tip. If the tip is at all jagged, it will probably not make a good holding pipet. If necessary, score the glass again a few millimeters from the tip and break it again to try to get an even tip. Make sure that the thin part of the pipet is at least 1 cm long.

3. Clamp the pipet into the DeFonbrune microforge in a vertical position, either using a Leitz instrument tube or else by inserting the pipet directly into the clamp on the microforge. Position the tip of the pipet so it is directly above the heating filament of the microforge. Heat up the filament and position it very close to the tip of the pipet until the glass begins to melt. Observe the inside diameter of the pipet as it melts by using an eyepiece with a micrometer and allow it to shrink to a diameter of about 15 μm. When it reaches the proper diameter, quickly turn off the current through the filament. If the filament is hot enough and is positioned close enough to the pipet, it should take only 30–60 seconds to complete the melting process. The hole in the end of the pipet should be straight, not pointing to one side, and the end of the pipet should be smooth and perpendicular to the long axis of the pipet (Fig. 57A). If it is crooked (Fig. 57B), uneven (Fig. 57C), or if the internal diameter is too small, try breaking a few millimeters off the end and melting it again.

4. Next, position the pipet horizontally in the microforge so that it is protruding over the filament, and move the filament so that it almost touches the pipet at a point 2–3 mm from the tip of the pipet. Carefully heat the filament until the pipet starts to soften at the point being heated, and allow it to bend under its

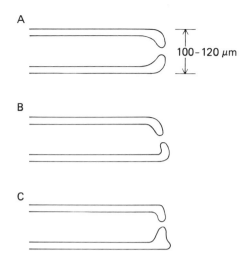

Figure 57 Holding pipet for microinjection. *A* shows the correct symmetrical shape of the tip, whereas *B* and *C* are incorrect.

own weight until it has bent by about 15° (see Fig. 58). Turn off the filament to stop the bending. The reason for introducing this bend is that it allows the end of the pipet to be positioned horizontally in the injection chamber (see Fig. 60). For some types of chambers, this bend is not necessary.

5. In the absence of a microforge, it is possible to fire-polish the end of the pipet by touching it to the flame of a microburner for the shortest possible time. The end will not melt as evenly as it does when prepared on the microforge, nor will it be possible to control accurately the diameter of the opening, so the majority of pipets will not be usable. A few, however, will be acceptable. Normally, one holding pipet lasts for a whole session of microinjection, but the pipet is not reused indefinitely because it tends to become dirty.

Making Injection Pipets

Use thin-walled glass capillary tubing, with an outside diameter compatible with the device in which the pipet will be inserted for injection (typically 1 mm or 1.5 mm). Tubing containing an internal glass filament (see Supplies, Section H) is convenient as it allows the pipet to be filled through the blunt end by capillary

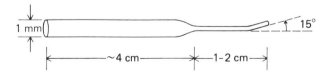

Figure 58 Holding pipet for microinjection. The pipet is shown after introducing a bend so that it can be used in the depression slide system illustrated in Figs. 60 and 61.

action. Start with a piece of tubing 10–15 cm long and use a mechanical puller to pull pipets with a suitable tip. This will require varying certain settings on the puller, such as filament temperature and strength of the pulling force. Keep track of the settings used to pull each pipet, in order to determine the best settings for future use. A good pipet for injecting mouse eggs should have a tip with a diameter of less than 1 μm; this is too small to be visualized clearly with a light microscope, so if you can see the hole in the end of the pipet when you look at it under the microscope, it is probably too large. Pipets with large tips are difficult to insert into the pronucleus, and tend to lyse many of the eggs. If the tip is too small, the hole will tend to clog very easily. The best way to learn what a good pipet looks like is to try to use various pipets and see which ones work best.

It is not necessary to modify the pipet tip in any way after pulling it on the mechanical puller. Some investigators prefer to grind the tip at an angle on an abrasive grinding wheel, or chip the tip against another piece of glass (under the microscope), yielding tips with larger openings. A larger opening, however, allows the fluid to flow more rapidly out of the pipet, making it more difficult to inject a controlled volume into the pronucleus. It also increases the tendency for culture medium to be accidently sucked back into the pipet, diluting the DNA solution.

Another important parameter is the degree of taper in the region near the tip of the pipet, which can be measured by the diameter of the pipet at a fixed distance from the tip. At a point of 50 μm from the tip of the pipet, the diameter should be about 10–15 μm or less. The distance between the shoulder and tip of the pipet is not critical, but should be at least 3–4 mm. If it is shorter, it may be difficult to position the tip near the bottom of the injection chamber, because the shoulder will interfere. The overall length of the pipet should be between 5 and 8 cm, for ease of handling. If too long, score with a diamond pencil and break at the desired location. Most mechanical pullers yield two pipets from every pull, and on some pullers the tips of the two pipets are similar so that both can be used. Glass capillaries can be used directly out of the package without cleaning or sterilization, when injected eggs are to be immediately transferred into the oviducts of foster mothers. If eggs are to be cultured for several days, it may be advisable to sterilize the capillaries before pulling the pipets, and then to store the pipets under sterile conditions. As pipets will be filled by dipping the blunt end into the DNA solution, avoid touching this portion of the pipets with fingers. Also avoid using powdered gloves to handle pipets, as the powder could clog the pipets. Pipets should be pulled the same day they are to be used, because old pipets are more likely to become clogged. They can be stored until use in petri dishes, supported by two ridges of plasticene or rolled-up masking tape.

The Microinjection Set-Up

Microscope

Microinjection is most easily performed using an inverted microscope with a fixed stage and image-erected optics. Two good microscopes for this purpose are the Zeiss IM35 and the Nikon Diaphot. Microinjection can also be performed using an upright microscope, but in this case the eggs must be handled in a hanging-drop culture chamber rather than in the culture chambers described below. A microscope specifically designed for microinjection work is marketed by Microinstruments (Oxford) Ltd. (see Equipment list). An image-erected microscope is not absolutely essential, as one can adapt to image-reversed optics, but it is easier to use. Similarly, a fixed-stage microscope (whose objectives move up and down for focusing) is not absolutely essential, but it is highly preferable. Microinjection of egg pronuclei is most easily performed at a magnification of 400–600×, usually achieved with a 40× objective and 10–15× eyepieces. In addition, a low-power objective, (4–10×) is used for moving eggs around in the injection chamber and for transferring eggs into and out of the chamber. For use with the inverted microscope, an objective must have a long enough working distance to be compatible with the thickness of the base of the injection chamber (typically a depression slide or plastic culture dish 0.5–1.5 mm thick). In addition, the condenser must have a long enough working distance to accomodate the injection chamber.

Mouse egg pronuclei can be visualized using standard bright-field optics, Nomarski differential interference contrast optics, or Hoffman modulation contrast optics. Nomarski or Hoffman optics are preferable because they make it easier to see the outline of the pronucleus and therefore easier to inject the pronucleus. Nomarski optics give somewhat finer resolution than Hoffman optics, but Hoffman has the advantage that it is considerably cheaper (see Supplies, Section H). Nomarski optics must be used with glass injection chambers, while Hoffman optics can be used with disposable plastic tissue culture dishes, which may be more convenient. Phase contrast is not helpful in visualizing pronuclei, and it is preferable to use a bright-field objective if Nomarski or Hoffman optics are not available. For the low-power objective, bright-field optics are sufficient.

Micromanipulators

The most commonly used micromanipulator for mouse egg micromanipulation is the Leitz micromanipulator. Micromanipulators made by Zeiss-Jena (East Germany), DéFonbrune (France), or Narashige (Japan) may also be acceptable. An important feature lacking in many other micromanipulators is the "joy stick," which allows simultaneous movement in two dimensions. Normally, one Leitz micromanipulator is mounted on each side of the microscope stage; one controls the holding pipet and the other the injection pipet. The micromanipulators must be firmly mounted at a suitable height relative to the stage and at a suitable distance from the stage (see Fig. 59). The Leitz micromanipulators can be used with some inverted microscopes (e.g., the Leitz Diavert) by simply mounting them on the same base as the microscope (such as the baseplate produced by Leitz for this purpose). The Nikon Diaphot and the Zeiss IM 35 have higher stages, and the micromanipulators must be mounted on elevated platforms to be used with these microscopes (see Appendix 1). It is important that the micromanipulators be attached to a common baseplate, either by bolts or by magnets, so that they do not move during use. The microscope may be attached to the same baseplate, or if it is heavy enough, it may simply rest on the baseplate under its own weight. If there are rubber feet on the microscope, they should be removed because they cause the microscope to vibrate when touched.

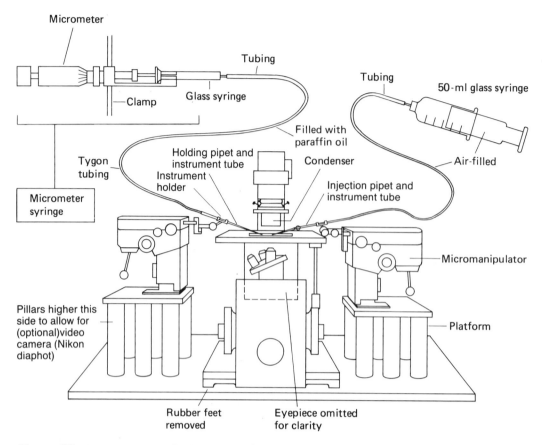

Figure 59 Arrangement of microscope, baseplate, and micromainipulators for the pro-nuclear injection method described in this manual. Details of the baseplate are given in Appendix 1, and sources of the apparatus are listed in this section.

Table

If the building is relatively free of vibration, it may be possible to set up the apparatus on a standard (low) lab bench, or on any sturdy table. Make sure that there is no source of occasional vibration (e.g., an ultracentrifuge) nearby, and that no one will be working on or bumping into the table during microinjection. If there is enough building vibration that the tip of the pipet vibrates when viewed under high magnification, additional measures must be taken to reduce vibration. One cheap solution that sometimes is sufficient is to mount the microscope and manipulators on a very heavy baseplate, and place some shock-absorbing material (e.g., high-density foam, tennis balls, motor scooter tires, or thick-walled rubber tubing) between the base plate and the table. If this does not solve the problem (and it usually will not if the vibration problem is severe), it may be necessary to place the apparatus on a special vibration-damping table, such as the Micro-G table (see equipment list).

Syringe to Control Holding Pipet

Connect a Leitz instrument tube, through a length of plastic tubing, to a syringe with a mechanism for fine movement (e.g., an Agla micrometer syringe; see Supplies, Section H). If the Agla syringe is used, a rubber band must be used to connect the plunger of the syringe to the micrometer assembly. Clamp the syringe in a position where it can be easily reached during micromanipulation. Fill the entire system with either light paraffin oil or Fluorinert FC77 (3M Company), taking care to eliminate any air bubbles. Fill the entire holding pipet with light paraffin oil or Fluorinert FC77, using a fine needle that can be inserted in the end of the pipet; insert the holding pipet into the instrument tube and wait until the oil or Fluorinert has stopped flowing out of the tip. Clamp the instrument tube with the pipet into the left-hand micromanipulator, and adjust the apparatus so that the tip of the holding pipet can reach into the injection chamber, as described below. View the tip of the pipet through the microscope, and adjust the syringe until the meniscus between the oil or Fluorinert and the culture medium is visible, close to the tip of the pipet.

Syringe to Control Injection Pipet

The Leitz instrument tube is connected, through a length of plastic (e.g., Tygon) tubing, to a 50-ml ground glass syringe (a plastic syringe is not suitable). The tip of the pipet is filled with about 1 μl of DNA solution (as described below) and the pipet is inserted into the instrument tube. Except for the DNA solution in the tip of the pipet, the entire system is filled with air. When pressure is exerted on the syringe, DNA is forced out of the tip of the pipet; when pressure on the syringe is released, the plunger returns to its neutral position and the flow of DNA ceases. The rate of flow is limited by the size of the hole in the needle and is easy to control as long as the hole is small. Negative pressure is never exerted because it might suck culture medium back into the pipet, diluting the DNA.

In an alternative set-up, also using variable positive pressure, the instrument tube is connected to a T-tube, one branch of which is connected to a compressed air outlet, and the other of which is connected to an open length of tubing. When air is flowing through the system, partial obstruction of the open branch (e.g., with a finger) creates pressure in the system, forcing the DNA to flow. This kind of system can also be controlled by a foot pedal, leaving both hands free.

Injection Chambers

1. The Petri dish chamber (see Fig. 60) is for use on an inverted microscope with bright-field or Hoffman modulation contrast optics, but not with Nomarski differential interference contrast optics. Use the inverted lid of a 90-mm plastic tissue culture dish (the lid has lower walls than the dish itself, allowing the instruments to be inserted at a lower angle). Make a large flat drop (about 5 mm square) of M2 culture medium about halfway between the center and the edge of the dish, and cover with paraffin oil. The drop should be as flat as possible because curved surfaces in the optical path cause distortion. The main advantage of this chamber is that it is disposable.

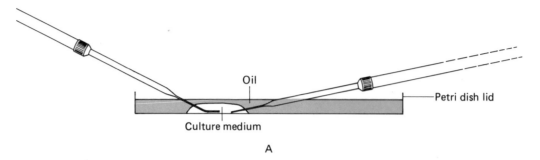

Figure 60   Petri dish injection chamber.

2. The depression slide chamber (Fig. 61) is for use on an inverted microscope with any optical system. Lightly siliconize a glass depression slide by dipping it in a 1% solution of dichlorodimethyl silane in chloroform. Rinse thoroughly in distilled water and autoclave. Place a large flat drop of M2 medium in the center of the depression, and cover with light paraffin oil. If the slide is too water-repellent, the drop will be too rounded; in this case, rinse the slide in chloroform or wash vigorously in dilute liquid detergent followed by rinsing to reduce the water repellency. If the slide is not siliconized at all, however, the drop will move around when the instruments are inserted, making it difficult to handle the eggs.

3. The cover slip chamber (not illustrated) is for use on an inverted microscope with any optical system. This chamber gives slightly better optical results than either of the above because it contains no curved surfaces, but it is more difficult to set up. Two rectangular bars of glass or plastic (about $3 \times 3 \times 10$ mm) are glued onto a standard glass slide. The slide is siliconized, and a rectangular cover slip, cut to about $10 \times 10 \times 20$ mm, is also siliconized. The tops of the two bars are lightly smeared with Vaseline or silicone grease. A small drop of M2 medium is placed in the center of the cover slip and another drop is placed on the slide, midway between the two bars. The cover slip is lowered (drop side down) onto the two bars, so that the two drops are lined up vertically. The entire space between the slide and the cover slip is filled with paraffin oil, and additional M2 is added to the two drops until they fuse, forming a continuous column of medium between the slide and the cover slip.

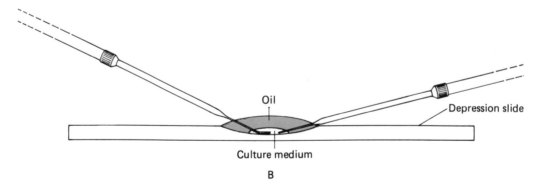

Figure 61   Depression slide injection chamber.

Injecting Mouse Egg Pronuclei

1. Set up an injection chamber, place it on the microscope stage, and focus on the bottom of the drop of medium, using a low-power (4–10×) objective. Insert the assembled holding pipet into the drop from the left side, and adjust it so that it lies horizontally on the bottom of the chamber.

2. Using an embryo pipet, transfer several mouse eggs into the chamber. Examine the eggs under high power, and make sure that the pronuclei are visible. If they are not visible, either the eggs are not fertilized (in this case the second polar body will not be visible), the eggs were too recently fertilized and the pronuclei have not yet formed, or the pronuclei have already broken down and the egg will soon divide. If the eggs have visible pronuclei, switch back to low power.

3. Fill the tip of an injection pipet with DNA by dipping the blunt end into the DNA solution. If capillaries containing an internal filament are used, the DNA solution will soon accumulate in the tip of the pipet. The pipet should be filled with DNA solution to a distance of several millimeters from the tip (approximately 0.5 µl). Clamp the pipet into the instrument tube connected to the injection syringe, attach the instrument tube to the micromanipulator, and insert the tip of the pipet into the drop of medium. The injection pipet is inserted at a slight angle (5–10°) to allow the tip to reach to the bottom of the chamber. The holding pipet and the injection pipet can both be used to push the eggs around in the chamber.

4. To make sure that the injection needle is not closed at the tip or clogged, place the tip of the pipet next to an egg in the same horizontal plane (i.e., in the same focal plane, on high power), and squeeze firmly on the injection syringe. You should be able to blow the egg away from the pipet with the stream of DNA; if you cannot, either the pipet has a closed tip or has already become clogged. Try a new injection pipet, and repeat this test.

5. To inject an egg, place the tip of the holding pipet next to the egg and suck it onto the end of the pipet. Focus the microscope to locate the pronuclei. A pronucleus can be most easily injected if it is located in the hemisphere of the egg closest to the injection pipet. In addition it should be as close as possible to the central axis of the holding pipet; if it is far from this axis, the egg will tend to rotate when the injection pipet is pushed toward the pronucleus. If necessary, reorient the egg to place one pronucleus in a better position. This is done by releasing it from the holding pipet, using the injection pipet and/or the holding pipet to rotate it slightly, and then sucking the egg back onto the holding pipet. When you are satisfied with the position of the egg, give the syringe controlling the holding pipet an extra twist to be sure the egg is firmly held. You should see the zona pellucida being pulled slightly into the opening of the pipet, but the egg itself should not be deformed. Either of the two pronuclei may be injected but the male pronucleus is usually larger and nearer the surface and therefore easier to "hit."

6. Next, refocus on the pronucleus to be injected, using the microscope focusing control. Place the tip of the injection pipet next to the egg and bring it into the

same focal plane using the vertical control on the micromanipulator. Squeeze firmly on the syringe before entering the egg to flush out any culture medium that may have entered the injection needle. Gently push the injection pipet through the zona and into the egg, toward the nucleus. Make sure both the tip of the pipet and the nucleus remain in focus; if the egg moves and the nucleus goes out of focus, the pipet may not hit the nucleus. Continue pushing the pipet forward until it looks like it is inside the nucleus. Avoid touching the nucleoli, as they are very sticky and will adhere to the pipet. When you think the tip of the pipet is inside the nucleus, squeeze on the syringe; if the nucleus swells obviously (see Fig. 62B), you have successfully injected it. If it does not swell, either the pipet has become clogged or your pipet has not punctured the egg plasma membrane. If you see a small round "bubble" forming around the tip of the pipet as you squeeze on the injection syringe (see Fig. 62A), the pipet has not punctured the plasma membrane. The membrane is very elastic, and it can apparently be pushed far back into the egg, even into the nucleus, without being pierced. In this case, try pushing the pipet *right through* the nucleus and out the other side; then pull back slightly so the tip is again inside the nucleus, continually squeezing on the syringe. This maneuver frequently gets the pipet through the plasma membrane and into the nucleus. Another sign that the pipet has actually pierced the membrane is that at the point of entry the membrane will be roughly perpendicular to the wall of the pipet (Fig. 50A), whereas if the membrane has not been pierced it will appear indented (Fig. 50B). When you see the nucleus swell, indicating that it has been injected, quickly pull the pipet back out of the egg. If you pull it out slowly, it will frequently remain attached to nuclear components (perhaps the nuclear membrane, or chromosomes) and you may pull out the nucleus with the pipet.

7. If you see cytoplasmic granules flowing out after removing the pipet, the egg may be starting to lyse; try another egg. If the egg appears to be intact, switch back to low power, place the injected egg to one side of the drop of medium, and pick up another egg. The same injection pipet can be used as long as it continues to inject successfully. You should switch to a new injection pipet if: (1) you are unable to get into the pronuclei of several eggs, even though you can clearly see the pronuclei; (2) two eggs in succession lyse immediately after injection; (3) the tip of the pipet becomes visibly "dirty," or something from the nucleus sticks on the pipet; (4) the tip of the pipet breaks and seems to be more than about 1 μm in diameter; or (5) the pipet clogs. On average, one injection pipet can be used to inject 5–10 eggs. When all the eggs in the chamber have been injected, put them back to a culture dish at 37°C, and transfer new eggs into the chamber. Transfer only as many eggs as you can inject in about 15 minutes, because it may be harmful for the eggs to be left for too long in M2 medium at room temperature.

8. When all the eggs have been injected, sort through them to separate healthy looking ones from lysed ones. A healthy egg will have a distinct outline and a perivitelline space between the egg and the zona pellucida (this can be easily seen under the dissecting microscope), while a lysed egg will fill the entire zona. Typically 50–80% of the eggs will survive the injection. Eggs that have survived microinjection can be: (1) transferred the same day into the oviduct of a 0.5-day p.c. pseudopregnant female; (2) cultured in vitro to the two-cell stage (overnight) and transferred into the oviduct of a 0.5-day p.c. pseudopregnant female;

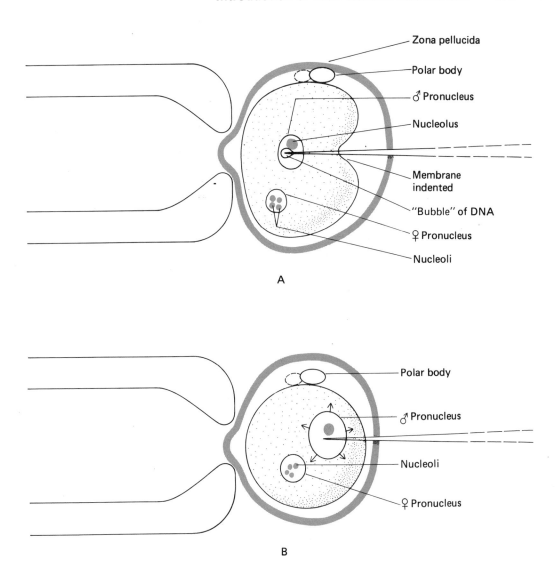

Zona pellucida

Polar body

♂ Pronucleus

Nucleolus

Membrane indented

"Bubble" of DNA

♀ Pronucleus

Nucleoli

A

Polar body

♂ Pronucleus

Nucleoli

♀ Pronucleus

B

Figure 62 Microinjection into the pronucleus. (*A*) Incomplete penetration of the egg plasma membrane. The plasma membrane is pushed into the pronucleus by the microinjection needle. The DNA solution expelled from the needle forms a visible bubble of plasma membrane within the pronucleus, but the pronucleus does not swell. When the needle is removed the DNA will flow back out of the egg. (*B*) Successful injection of the pronucleus. The plasma membrane has been pierced, and the egg returns to a spherical shape. The tip of the needle remains inside the pronucleus, and DNA expelled from the needle causes the pronucleus to swell visibly.

(3) cultured in vitro to the blastocyst stage and transferred into the uterus of a 2.5-day p.c. pseudopregnant female. The first two options will probably yield the highest ratio of survival to term, because they avoid extended in vitro culture. Approximately 10–30% of microinjected eggs transferred into the oviduct will develop to term. To avoid very small litters, which may not be cared for by the foster mother, it is advisable to transfer 20–30 microinjected eggs into each pseudopregnant recipient.

A Protocol for Isolating High-molecular-weight DNA from Mouse Tails

This protocol* describes the isolation of DNA from mouse tails using adaptations of various standard methods for isolating high-molecular-weight DNA. Tails from 2-week-old mice are routinely used, though mice up to 7 months have been used to produce high-molecular-weight DNA of similar yield and quality. About 50–200 μg of high-molecular weight DNA is obtained from 1–2 cm of a tail of a 2- to 3-week-old mouse. The OD 260/280 ratio of the DNA is 1.8 ± 0.1, and the DNA is of reasonably high molecular weight ($M_r > 30$ kb), can be digested readily with *Eco*RI, *Bam*, *Bgl*II, *Taq*, *Sac*, *Kpn*, and *Pvu*II (*Hind*III and *Xba* do not digest this DNA well), and has produced excellent Southern blots. The variation in yield is substantial, so it is advisable to determine the DNA concentration using UV absorbance or dye-fluorescence prior to DNA blotting.

PROCEDURE

1. Cut 1.3 cm of tail into a 1.5-ml microfuge tube containing 700 μl of (50 mM Tris, pH 8, 100 mM EDTA, 100 mM NaCl, 1% SDS). Some bleeding will occur, but no special treatment is required. Earmark the mouse appropriately. (Mouse should be >2 weeks old.)

2. Mince tail using a pair of small, very sharp, stainless steel scissors that will fit the bottom of the tube.

3. Add 35 μl of a 10 mg/ml solution of Proteinase K dissolved in H_2O (Merck 24568, EM Biochemicals, available from American Scientific Products).

4. Incubate at 55°C for 8–18 hours (overnight). Invert tube occasionally to mix, or place on a rocking platform.

5. Remove tubes from 55°C. Add 20 μl of RNase (Worthington #5679, 13 μg/ml). Incubate at 37°C for 1–2 hours.

6. Fill tube with phenol. Place on vertical rotator (horizontal axis of rotation) for 15–20 minutes.

7. Separate phases in a microfuge (10- to 15-min spin).

8. Transfer aqueous phase, along with interface, to a fresh microfuge tube.

---

*Information supplied by Dr. Douglas Hanahan, Cold Spring Harbor Laboratory, Cold Spring Harbor, New York 11724.

9. Fill tube with phenol/chloroform (1:1); rotate 5–10 minutes.

10. Separate phases in a microfuge (5-min spin).

11. Transfer aqueous phase, again with any interface, to a fresh tube.

12. Fill tube with chloroform/isoamyl alcohol (24:1). Rotate 5 minutes, spin 5 minutes.

13. Transfer aqueous phase to a fresh tube, leaving any interface behind. Fill tube with isopropanol. Invert tube several times, until stringy precipitate forms.

14. Remove stringy precipitate by touching it to the sealed end of a 100-μl micropipet. (The 100-μl pipet is flamed at one end to seal it. A small piece of tape is placed on the other end, with the mouse number on it.)

15. Dip spooled DNA into 70% ethanol.

16. Dip spooled DNA next in 100% ethanol.

17. Allow DNA to dry. The opposite end of the 100-μl micropipets can be inserted into styrofoam leaving the DNA tips safely separated in an ordered array.

18. Wearing gloves, score the micropipet with a diamond pencil 3 cm from the spooled end.

19. Break off spooled tip and drop it (DNA tip down) into a 1.5-ml microfuge tube containing 500 μl of TE buffer (10 mM Tris, pH 7.4, 1 mM EDTA).

20. Place tubes in a rack on a horizontal rotator at 30–60 rpm overnight (vertical axis of rotation).

21. Remove glass tip using alcohol-flamed forceps.

22. Place tubes on a vertical rotator for a few hours.

23. Remove aliquot to determine concentration (e.g., 5% of the DNA).

24. Subject 2% of DNA solution to 0.7% agarose gel electrophoresis to assess quality of the DNA. Phage DNA digested with *Hind*III provides reasonably sized markers when electrophoresed such that the 23-kb band is midway through the gel.

An Alternative Protocol for Isolating High-molecular-weight DNA from Mouse Tails

This protocol is somewhat simpler, and also yields a similar amount of high-molecular-weight DNA that can be digested with most restriction enzymes.

1. Cut 1–2 cm of tail and place in a 1.5-ml microfuge tube. Mincing the tail is not necessary.

2. Add to the tube 0.7 ml of 50 mM Tris (pH 8), 100 mM EDTA, 0.5% SDS. Add 35 μl of a 10 mg/ml solution of Proteinase K.

3. Incubate at 55 °C overnight, on a rocking platform (Clay Adams aliquot mixer).

4. Remove tubes from 55 °C. Add 0.7 ml of phenol (equilibrated with Tris [pH 8]; see Maniatis et al. 1982). Close tube and shake *vigorously* for 3 minutes, so that phases mix completely.

5. Centrifuge in a microfuge for 3 minutes, so that phases separate.

6. Transfer aqueous phase to a fresh tube, being careful not to pick up phenol or material at interface.

7. Add 0.7 ml of phenol/chloroform (1:1), shake vigorously 2 min, and centrifuge 2 minutes.

8. Again remove aqueous phase, avoiding interface, and transfer to a 1.8-ml microfuge tube (Sarstadt).

9. Add 70 μl of 3 M sodium acetate (pH 6), and 0.7 ml of 100% ethanol at room temperature. Shake to mix thoroughly. DNA should immediately form a stringy precipitate. A sodium acetate solution with a pH lower than 6 will cause the EDTA to precipitate.

10. Spin in microfuge for 30 seconds to pellet DNA. Remove and discard as much as possible of ethanol supernatant.

11. Add 1 ml of 70% ethanol (room temperature) to tube, and vortex or shake vigorously to wash DNA. This step is essential to remove traces of SDS and phenol.

12. Centrifuge in microfuge for 1 minute at room temperature. Remove as much ethanol supernatant as possible. Dry DNA briefly in vacuo.

13. Add 0.1 ml of (10 mM Tris [pH 8] 1 mM EDTA) to tube. Leave at room temperature overnight to dissolve. If necessary, DNA can be dissolved more quickly by heating for 5–10 minutes at 65 °C. Use 10–20 μl of each DNA preparation for Southern blot analysis.

14. DNA prepared in this manner will contain substantial amounts of RNA, but this does not interfere with restriction enzyme digestion or Southern blot analysis. When performing retriction enzyme digestion, add 5 μg of DNase-free RNase A to each sample along with the restriction enzyme. Digestion with certain restriction enzymes may also be aided by adding 4 mM spermidine to the reaction (Bouche 1981).

Biopsies: Partial Hepatectomy, Splenectomy, Nephrectomy, and Tail Bleeding

Partial Hepatectomy

EQUIPMENT

> Avertin (anesthetic)
> Animal balance
> 1-cc syringe
> 26-gauge 1.3-cm needle
> Animal clippers
> 70% ethanol
> Tissues
> Dissection scissors
> Two watchmaker's #5 forceps
> One blunt forceps
> Surgical silk suture (5-0)
> Surgical needle
> Alcohol burner
> Fiber optic illuminator
> Wound clips and applicator

PROCEDURE

Note: This procedure can be performed by one person, but is more easily performed by two.

1. Anesthetize the mouse by injecting it i.p. with 0.015–0.017 ml of 2.5% Avertin per gram of body weight. Shave the upper abdomen and wipe the skin with 70% ethanol to remove loose hairs. Flame instruments with an alcohol burner.

2. With dissection scissors, make a transverse cut about 1.0 cm long in the skin just below the sternum. Wipe again with 70% ethanol. Pick up the body wall with watchmaker's forceps and make a transverse cut, also just below the sternum. There are two major blood vessels, about 1 cm apart, which run longitudinally in the body wall, on either side of the midline. Avoid cutting through them, as considerable bleeding will result (this will not endanger the animal's life, but it will obscure the incision).

3. Before starting the operation, cover the ends of the blunt forceps with a single layer of plastic or paper tape. This will make it easier to get a grip on the liver. Reach inside the body wall with one or two watchmaker's forceps, and pick up the edge of the top lobe of the liver. Gently bring the liver lobe outside the body wall, and then grasp it with the taped end of the blunt forceps, holding the forceps horizontally. Ease the lobe of liver farther out until about two-thirds of it is exposed. Have second person thread a piece of surgical silk underneath the lobe and tie it off to constrict the blood vessels and reduce bleeding when the liver is cut. While the second person holds onto the thread to keep the liver

from slipping back into the body, the first person cuts off the lobe with a scissors and rinses off the blood in a dish of cold PBS. The lobe can now be frozen in liquid N$_2$, or used immediately for DNA or RNA preps.

4. Sew up the body wall and then clip the skin closed.

Splenectomy

EQUIPMENT

Same as for Partial Hepatectomy

PROCEDURE

1. Anesthetize the mouse, shave the fur on the left side of the lower back, and wipe with 70% ethanol. Flame instruments with alcohol burner.

2. With dissection scissors, make a small transverse incision in the skin, just to the left of the spinal cord at the level of the last rib. Wipe with 70% ethanol. Make a transverse cut in the body wall and pull out the spleen using a blunt forceps. Cut the gastro-splenic ligament, which is at one end of the spleen. A major artery and vein run together through the fatty tissue to the middle of the spleen. Tie off these vessels together with surgical silk. Cut the vessels distal to the knot and remove the entire spleen. Sew up the body wall and close the skin with wound clips.

Nephrectomy

EQUIPMENT

Same as above.

PROCEDURE

1. Anesthetize the mouse. Shave the fur on the left side at the level of the middle of the back, wipe with 70% ethanol, and flame instruments with the alcohol burner.

2. With the dissection scissors, make a 0.5- to 1.0-cm longitudinal incision in the skin, just to the left of the spine, at the level of the last rib. Wipe with 70% ethanol. Make 0.5-cm longitudinal cut in the body wall and with a needle draw a piece of silk suture through the dorsal body wall.

3. Locate the kidney and gently squeeze the mouse to push the kidney through the body wall. Alternatively, try to pull the kidney out by grasping the fatty tissue at the base of the kidney with a blunt forceps. Pull a thread underneath

the kidney and tie off the renal artery, renal vein, and ureter with two knots. Cut off most of the kidney with the scissors, leaving a small amount behind to keep the knot from slipping off.

4. Sew up the body wall and close the skin with wound clips.

Tail Bleeding

EQUIPMENT

> Lamp with 100-watt bulb
> Razor blade
> Box or large test tube, with hole for the tail and air holes, to re-
>   strain the mouse
> Microfuge tube or 5-ml Falcon tube
> Heparin (optional)
> 70% ethanol

PROCEDURE

1. Place the mouse in the box with the tail sticking out and wipe the tail with 70% ethanol. Warm the tail close to the lamp for about 5 minutes to increase the blood flow. If you wish to inhibit clotting, place a small drop of heparin (about 1000 U/ml) in the tube and shake the tube to coat the inside surface with heparin.

2. With a razor blade (or a sharp scissors) cut off about 1 cm from the tip of the tail. Continue to hold the tail under the lamp, and gently stroke it toward the cut to promote bleeding. Collect the drops of blood in the heparin-treated tube. About 0.5 ml can be collected from a 20-g mouse without endangering its life.

3. If the blood clots and stops flowing, cut off another short segment of tail. Squeeze the tip of the tail, if necessary, to stop the bleeding.

An alternative procedure for obtaining 100–200 μl of blood is the following:

1. Etherize the mouse by placing it on ether-saturated chamber until it is unconscious.

2. Grasp the mouse firmly by the scruff of the neck, and cut off the end of the tail.

3. Quickly insert the tail into a heparin-treated tube and hold it there until the desired volume of blood is collected. The mouse will jerk its tail when it starts to regain consciousness, so hold the tail carefully in the tube to avoid losing the blood.

Analysis of Foreign DNA in Transgenic Mice

Animals born from injected embryos may be initially identified either by dot blot analysis (Kafatos et al. 1979) or by restriction enzyme digestion and Southern blot analysis (Southern 1975). While dot blot analysis is simpler and faster, we prefer to screen for founder transgenic mice by Southern blot analysis for several reasons. First, mice carrying very low numbers of copies of the injected gene may be difficult to detect by dot blot analysis, unless the background hybridization is very low. If a founder mouse has a low copy number and is mosaic, it could easily be missed on a dot blot. In contrast, the presence of bands of the predicted size(s) on a Southern blot clearly indicates a positive transgenic mouse, and under appropriate conditions less than one copy per cell of a gene can be detected. Second, contamination of a mouse DNA sample with plasmid DNA (which can occasionally happen despite extreme efforts to avoid it) could cause a false positive on a dot blot; on a Southern blot, a contaminating plasmid is likely to produce bands that differ in size from the expected bands. Third, a Southern blot provides immediate information about the structure and integrity of the inserted DNA sequences.

The choice of restriction enzymes for the analysis of the integrated DNA is important, as is the choice of the hybridization probe. Normally, the injected DNA fragment, or a plasmid containing the fragment, can simply be labeled by nick-translation (Maniatis et al. 1982) and used as the probe. If the fragment contains a repetitive sequence that would hybridize to related sequences in the host genome, then a subfragment, free of repetititve sequences, may have to be used as a probe instead.

When choosing a restriction enzyme, keep in mind that in most cases the injected DNA fragment will integrate in a head-to-tail array, as discussed above. If one selects a restriction enzyme that does not cut the injected DNA, the fragment produced will be larger than the entire array, assuming that the foreign DNA has integrated as a simple array and has not undergone any other rearrangments. It is preferable to use initially an enzyme that will yield bands of predictable sizes. An enzyme that cuts once in each injected molecule will yield a band of the same length as the injected fragment, if a head-to-tail array has been formed. It should also yield "junction fragments" of novel lengths from the two ends of the array. If a single copy has been integrated, such an enzyme should yield only two fragments of novel length. Regardless of whether one copy or a head-to-tail array has integrated, an enzyme that cuts twice in the injected DNA molecule will yield a predictable band on a Southern blot, derived from an internal fragment in the injected DNA molecule. In the case of a head-to-tail array, a second predictable fragment representing the junction between adjacent members of the array will be generated.

To determine the number of copies of a foreign DNA molecule that are integrated in the genome of a transgenic mouse, it is first necessary to determine accurately the concentration of a genomic DNA preparation from that mouse. Many DNA preparations, such as those purified from tail skin by the procedure described above, may be heavily contaminated with RNA. Therefore, the absorbance at 260 nm will not accurately reflect the DNA concentration. Instead, a fluorimetric DNA determination must be performed (e.g., Thomas and Farquhar 1978 or Labarca and Paigen 1980). Varying amounts of genomic DNA (typically 0.5, 1, 2, and 4 μg) are spotted onto a nitrocelluose filter (Kafatos et al. 1979); in addition, accurately known amounts of the injected DNA fragment are mixed with normal mouse DNA and spotted onto the same filter to produce a standard curve. One copy of a 5- to 10-kb sequence per diploid mouse genome (6×10^9 bp) is roughly one part in one million; therefore, the amount of the pure DNA fragment used for the standard curve should cover a range of approximately 1–100 pg.

The dot blot is hybridized using the injected DNA fragment or plasmid as a probe, and the hybridization to each dot is quantitated by scintillation counting, or by autoradiography and densitometry. If the amount of hybridization is linear with the amount of DNA spotted, for both the transgenic mouse DNA curve and the standard curve, the copy number can be calculated from the slopes of the two lines. If the number of copies integrated is very high, the curve generated with transgenic mouse DNA may not be linear because the amount of probe added to the hybridization may not be sufficient to saturate complementary sequences in the mouse DNA. In this case, the procedure must be repeated using either lower amounts of transgenic mouse DNA, or a higher concentration of probe, or both. Gene copy number can also be estimated by including standard amounts of the injected gene in parallel lanes on a Southern blot of transgenic mouse DNAs.

Note that the apparent copy number measured by such a procedure reflects both the actual copy number per diploid genome and the fraction of cells containing the foreign DNA. If a founder mouse is mosaic and some cells lack the injected gene, the copy number will be underestimated. For this reason, it is preferable to determine gene copy number using transgenic progeny rather than the founder.

Occasionally a founder transgenic mouse will contain foreign DNA integrated at two different loci (Lacy et al. 1983; Wagner et al. 1983). Although this can be detected directly by performing in situ hybridization to metaphase chromosome spreads (see below), it will also become apparent when progeny are analyzed, because the two loci will segregate. Since the different integration loci will usually contain different numbers of copies of the injected DNA, the inheritance of different numbers of copies by progeny frequently provides the first evidence for multiple integration sites in the founder. In addition, each integration event will generate distinct junction fragments, and these may be seen to segregate when progeny DNAs are analyzed by Southern blotting.

Identification of Homozygous Transgenic Mice

Three methods have been used to identify mice that are homozygous for a foreign DNA integration site in the offspring from an intercross between two heterozygous transgenic mice. First, foreign DNA copy number can be quantitated by dot blot analysis, as described above: homozygotes will contain double the number of copies found in heterozygotes. A somewhat more reliable variation of this method involves measuring the relative quantity of foreign DNA in each mouse, using an endogenous gene as an internal standard. The use of an internal standard controls for any errors in measuring DNA concentration; in fact, this method eliminates the need to determine accurately DNA concentrations. For this purpose, duplicate dot blots are prepared, each containing a series of dilutions of DNA from the progeny to be tested, as well as from one or two known heterozygous mice. One blot is hybridized with a probe for an endogenous gene. The endogenous gene used as an internal standard should have a similar reiteration frequency to the foreign gene, so that similar amounts of hybridization (and linear curves; see above) will be obtained with each probe. For mice carrying 1–10 copies of a foreign gene, any single-copy mouse gene can be used as an internal standard. For mice carrying 10–100 copies of a foreign gene, a convenient internal standard is the mouse major urinary protein (MUP) gene (20–30 copies per haploid genome; Derman 1981; Derman et al. 1981). For the range of 50 to several hundred copies, a probe for a ribosomal RNA gene (Arnheim et al. 1982) works well. For each mouse, the slope of the curve (cpm hybridized/μg DNA) obtained with the foreign gene probe is divided by the slope obtained with the endogenous probe. The ratio for a homozygote should be twice the ratio for a heterozygote.

A second method is to perform a Southern blot analysis using a probe (or mixture of probes) that will yield distinct bands for both the foreign DNA and an endogenous gene (Wagner et al. 1983). In this method, the intensity of a band representing the foreign DNA is compared with that of a band representing an endogenous gene, using densitometry of the autoradiogram. Although this method is simpler than the dot blot method, it is subject to error due to factors such as nonuniform transfer of DNA to the filter, or incomplete digestion with the restriction enzyme. In addition, for transgenic mice with large numbers of copies of the foreign DNA, it may be difficult to find an endogenous gene that is similarly reiterated and that also yields a discrete band on a Southern blot. If one intends to mate two homozygous transgenic mice to generate a permanent homozygous line, it is advisable first to confirm the homozygosity of each animal by genetic means, because quantitative methods for identifying homozygotes can sometimes produce erroneous results. This is done by outcrossing each presumptive homozygous mouse to a nontransgenic mouse and checking for 100% transmission of the foreign DNA to the progeny.

A more definitive means of identifying homozygotes (short of the genetic method) involves the use of a host DNA sequence at the integration site as a probe (Jaenisch et al. 1983). This, of course, can only be done if the host DNA flanking the foreign DNA insertion has been cloned. When used to probe a Southern blot of DNA prepared from heterozygous mice, such a host DNA sequence will detect two bands: one containing the foreign DNA insert and the other containing the preintegration site on the homologous chromosome. On a similar Southern blot, mice that are homozygous for the foreign DNA insertion will lack the band representing the preintegration site, eliminating the need for quantitation.

DETERMINATION OF SEX UTILIZING RETROVIRAL DNA PROBES

The sex of adult mice is generally determined by examining external genitalia, and that of embryos by dissection to reveal sexual organs. However, in cases where it is not possible to use the normal criteria, male and female mouse embryos and tissue samples can often be discriminated by another method, using retroviral DNA probes, based on a report by Phillips et al. (1982). Nearly all species of *Mus* tested have high levels of endogenous type C-I retrovirus-related sequences widely distributed in their genomes. One of the type C-I family members, M720, has related sequences that are repeated ~100 times on the mouse Y chromosome. Therefore, when mouse genomic DNA is probed with M720, male-specific M720-related *Eco*RI fragments (14.5, 11.8, 7.5, and 4.3 kb) are easily detected among the autosomal and X-related sequences, and serve as the basis for determining the original sex of the animal or embryo.

In practice, this has proved to be an effective and sensitive method (Krumlauf et al. 1986) because the sequences are so highly repeated and very little DNA is required (0.1–1.0 μg). The M720 probes have been successfully used in BALB/c, BALB/cJ, C57BL, AKR, NZB, DBA, SM, S3L, and C3H strains of *Mus musculus*. The original paper utilized the entire 8.5-kb cloned M720 sequence in pBR322 to probe *Eco*RI-digested mouse DNA. However, the use of two internal *Hin*dIII fragments of the M720 clone (1.5 and 1.9 kb) produces a pattern with fewer background bands that is easy to read. In addition to *Eco*RI, the male-specific bands can also be detected with *Hin*dIII, *Pst*, and *Xho*I, and other enzymes may also be useful.

In the future it may be possible to use Y-specific repeated sequences recently isolated by Dr. Eva Eicher to determine sex by probing DNA dot blots or by in situ hybridization of very early embryos.

Information supplied by Dr. Robb Krumlauf, National Institute for Medical Research, Mill Hill, London NW7 1AA, England.

RETROVIRAL INFECTION OF PREIMPLANTATION EMBRYOS

MATERIALS

Retrovirus-producing cell line, subcultured into 60-mm tissue culture dishes

Polybrene, stock solution of 2 mg/ml

Alpha-modified minimal essential medium (MEM) plus 10% FBS

Acid Tyrode solution (see Section G)

PROCEDURES*

1. Collect eight-cell embryos on day 2 p.c. by flushing the oviducts with M2 (Section F). Remove zonae with acid Tyrode solution (Section G) and wash the embryos.

2. Wash a dish of fibroblasts producing retroviruses free of any selection medium and replace the medium with alpha-MEM plus 10 μg/ml Polybrene. Cultures should be subconfluent and still in log phase of growth for maximum virus production.

3. Zona-free embryos are added to the culture dish and the embryos and cells are cocultivated for 24 hours. After this time, which should be at the late morula–early blastocyst stage, remove the embryos from contact with the cells and transfer them to the uterus of day 2.5 p.c. pseudopregnant females (Section C).

4. Filter the medium overlying the cells and freeze at $-20\,^\circ$C for later estimation of virus titer. Analyze results during pregnancy by isolating tissues and extracting DNA and RNA, or allow the fetuses to proceed to term. Live mice can be screened for viral insertion by Southern blots of tail DNA, and germ line transmission can be assessed after crossing to uninfected mice.

Comments

1. The titer of the virus stock must be high (5×10^5 or higher) in order to obtain high infection rates. The percentage of offspring that contain the viral sequences has been reported to vary between 10 and 50% in different labs.

2. Polybrene may help to increase infection rate and is not toxic to embryos at the doses used. In one set of experiments we obtained a rate of 18% incorporation without Polybrene and 37% with Polybrene.

*Information provided by D.J. Rossant, A. Bernstein, and H. Tanenbaum, Department of Research, Mount Sinai Hospital, Ontario, Canada.

3. Infection at the eight-cell stage results in mosaic mice which do not contain the virus in all their cells. Sometimes multiple inserts can be detected in offspring, presumably due to viral insertion into more than one cell. The consequences of this mosaic incorporation are: (1) that detection of viral sequences in the founder mice may be difficult if an infected cell contributes to only a small part of the fetus, (2) that integration into the germ line may not occur in all animals, and (3) that the effect of any genes carried on the retrovirus may be difficult to assess. It may be possible to infect earlier embryos by mixing embryo culture and tissue culture media during the period of cocultivation, but nonmosaic integration of retroviruses has not yet been reported.

IN UTERO INJECTION OF VIRUSES AND CELLS

This procedure* allows the efficient introduction of virus into cells of all somatic tissues by direct injection of virus (Jaenisch 1980) or virus-producing cells (Stuhlmann et al. 1984). Chimeric mice can be generated by microinjection of normal cells into midgestation embryos (hemopoietic cells, Fleischman and Mintz 1979; neural crest cells, Jaenisch 1985).

EQUIPMENT

Anesthetic (see below)
Glass injection pipet made from thin-walled Leitz glass tubing
Microforge with fine filament
Stereomicroscope
Equipment for opening and closing abdomen wall (see Vasectomy, for example)

Recipient are females between 7 and 10 days of pregnancy.

*Information provided by Dr. Heidi Stuhlman, Whitehead Institute, Cambridge, Massachusetts 02139.

PROCEDURE

Some anesthetics result in a high proportion of spontaneous abortion after the operation. The drug Hypnodil (Methomidat-HCl; Janssen Chemie Gmbtt.) at a dose of 20–25 U/g body weight has been found to give good results, supplemented with ether if necessary. The ether can be administered by placing a glass tube containing ether-soaked cotton wool just over the head of the mouse.

1. An injection needle is pulled to give a tip of about 20 μm diameter. The end is sealed in a microforge so that it is sturdy yet sharp enough to puncture the uterine wall. A hole is made near the tip of the needle by connecting the pipet to a source of compressed air, and bringing the heated filament to the microforge close to the side of the pipet (Fig. 63A). When a hole forms in the side of the pipet, the current to the filament is quickly switched off.

2. Open the abdomen of the mouse and gently pull out the guts and then the two horns of the uterus with blunt forceps. Spread out the uterus as shown in Figure 63B. Locate the placenta within each conceptus and hold the tip of the conceptus at the end furthest away from the mesometrium (membrane attaching uterus to body wall) with watchmaker's forceps.

3. Inject approximately 0.5 μl of fluid containing virus or cells anywhere in the antimesometrial one-third of the embryo beneath the placenta (Fig. 63C). Cells are kept on ice (to prevent clumping) at a concentration of about 10^3 to 5×10^3/ml and are sucked up into the pipet shortly before injection.

4. Return the uterus and guts gently to the abdomen, and close the body wall with stitches and the skin with wound clips.

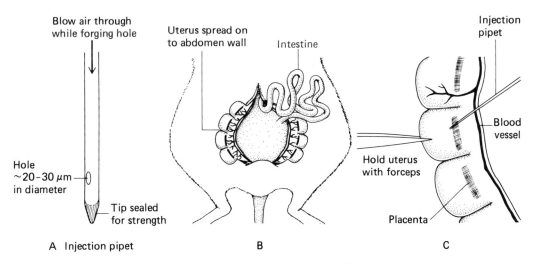

Figure 63 Method for in vitro injection of viruses and cells.

INJECTION OF CELLS INTO THE BLASTOCYST

The method for microsurgical injection of blastocyst-stage embryos was first described by Gardner (1968, 1971). The technique described here* represents a simplified version of that devised by Gardner in that only two micromanipulation instruments are employed: a holding pipet for immobilization of the blastocyst and an injection pipet for the introduction of cells into the blastocoel cavity. The basic microscope/micromanipulator set-up described in Microinjection of DNA into Pronuclei can be easily modified for blastocyst injection.

EQUIPMENT

> Right-hand and left-hand Leitz micromanipulators with single instrument holders
> Inverted fixed-stage microscope (e.g., Nikon Diaphot, Leitz Divert, or Zeiss IM35)
> Cooling stage and injection chamber for microscope (see Fig. 64)
> 2 micrometer syringes
> Pipet puller
> Microforge (e.g., DeFonbrune)
> Light paraffin oil
> Borosilicate glass tubing, 0.8 mm i.d., 1.0 mm o.d. (Corning Glass, Corning, New York.)

*Information provided by Drs. Liz Robertson and Allan Bradley, Department of Genetics, University of Cambridge.

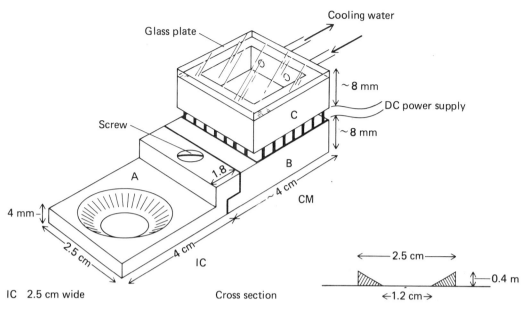

Figure 64   Design for a cooled injection chamber. The unit is constructed from aluminum and is in two sections—the injection chamber (IC) and the cooling module (CM). The chamber (A) is constructed by milling a circular depression into the center of the aluminum block. The lower surface is sealed with a glass plate. The unit is positioned on the microscope stage so that the chamber is aligned between the objective and the condenser. Cooling of the injection chamber section is provided via the cooling module. The DC current-regulated cooling unit is sandwiched between the two blocks of aluminum. The lower block (B) is cooled and is attached to the injection chamber. The upper block (C) is warmed by the action of the cooling unit. The heat is removed by a through current of circulating cold water. The injection chamber is calibrated by altering the DC current to give the desired working temperature.

Preparation of the Microinstruments

The most important feature of successful blastocyst injections is the quality of the tools used, particularly the injection pipet.

1. To prepare the injection pipet, pull glass capillary tubing on a mechanical pipet puller with the settings appropriately adjusted to give a long, gradually tapering end. A distance between the shoulder and the needle point of 1.25–1.5 cm is ideal. Then snap the pulled needle to give a sharp, beveled end. This is most reliably achieved by breaking the pulled needles, supported on a "spongy" silicone rubber pad, at the correct diameter (15 μm) using the edge of a sharp scalpel blade. This procedure can be undertaken by hand under a binocular dissection microscope (20×). In general, about half of the broken needles are sufficiently sharp and of the correct dimensions. The type of glass used may be important. Borosilicate glass tubing 0.8 mm i.d. and 1.0 mm o.d. is recommended.

2. To prepare holding pipets, use the same type of glass capillary. Break a pulled capillary on the microforge at a diameter of about 90–100 μm. To do this, place the pulled glass, at the point at which it reaches the required diameter, just in contact with the glass bead on the filament and heat until the two glass surfaces fuse. Then turn off the filament and the capillary will break cleanly at right angles to the point of contact (see Fig. 65). Heat-polish the end using a hot filament until the internal diameter is in the region of 15–20 μm. (For an alternative method of preparing the holding pipet, see Microinjection of DNA into Pronuclei.)

Both the injection and holding pipets are bent on the microforge, at an angle of approximately 30°, to ensure that their ends are held parallel to the bottom surface of the injection chamber.

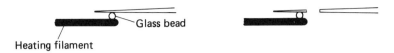

Figure 65 Breaking tip of micropipet using DeFonbrune micro-forge. Prepare a "glass anvil" on the tip of the thick platinum filament by fusing a small fragment of glass to the filament and allowing it to melt to a rounded ball. The glass anvil will remain in place as long as the filament is not heated to too high a temperature. Position micropipet above glass anvil and heat filament to a temperature at which the pipet will just barely fuse to the glass anvil, but will not be distorted (this temperature must be determined empirically). Touch the pipet to the glass anvil until it fuses, then turn off heat to the filament. As the filament cools and contracts slightly, it will break the pipet at the point of contact, leaving a flush end.

Micromanipulator Assembly

The injections are carried out using a Leitz double micromanipulator assembly with a fixed-stage, inverted microscope (see Microinjection of DNA into Pronuclei). An overall magnification of 50× is suitable for low-power operations, such as setting up the instruments, whereas a high-power, 200× phase-contrast overall magnification is required for the actual injection procedure.

The instrument holders are connected to controlling micrometer syringes using small-diameter, thick-walled, flexible tubing filled with light paraffin oil. The holding pipet is held in the left-hand micromanipulator with the micrometer syringe positioned within easy reach of the operator's right hand. The injection pipet, which is normally back-filled with low-viscosity silicone oil, is placed in the right-hand micromanipulator, with care being taken to remove any air bubbles from the system. The micrometer syringe controlling the injection pipet is positioned at the left-hand side. The instruments are aligned within the injection chamber so that the ends move in parallel to the bottom surface. All the injections are carried out in HEPES-buffered medium overlaid with paraffin oil. The injection medium consists of DMEM + 10% fetal bovine serum + 20 mM HEPES. Injections are carried out at 10°C. This cooling is important because it not only gives the cell membranes a degree of rigidity but also prevents "stickiness" which would otherwise necessitate the frequent changing of the injection needle. Figure 66A illustrates the micromanipulation set-up used. The orientation of the bevel of the injection pipet is important. Injections are carried out under a standard phase-contrast microscope so that the orientation of the needle end has to be carried out in two dimensions and the only information as to the exact position of the needle in relation to the three-dimensional blastocyst is provided by the fine focus. It is therefore an advantage to arrange the instruments as shown in Figure 66B. The correct blastocyst orientation is also shown in Figure 66B. The blastocyst is held in position by applying suction from the holding pipet and is prevented from moving in the vertical plane by lowering the embryo until it rests against the bottom surface of the injection chamber.

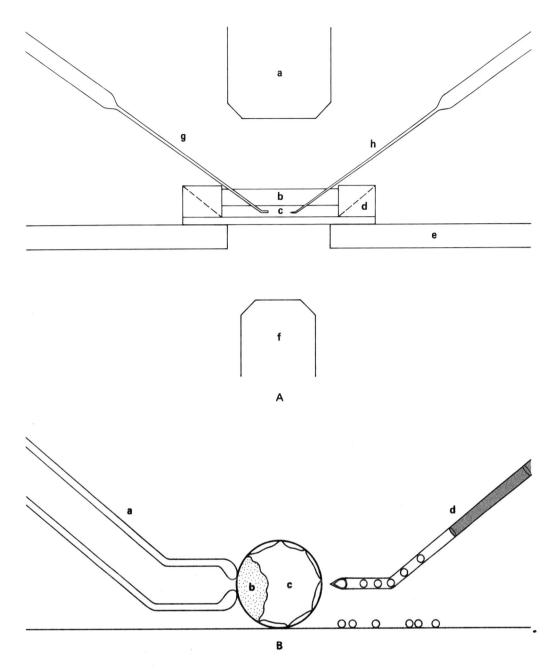

Figure 66 (*A*) Micromanipulator setup used for blastocyst injection. a. Microscope condenser; b. layer of paraffin oil; c. HEPES-buffered culture medium; d. aluminum injection chamber; e. microscope state; f. microscope objective; g. holding pipe held in left-hand manipulator; h. injection needle held in right-hand manipulator. (*B*) Orientation of instruments and embryo in the injection chamber (side view). a. Holding pipe; b. inner cell mass; c. blastocoel cavity; d. injection needle containing cells.

Blastocyst Injection

1. Transfer expanded blastocysts in groups of about 10 into the precooled injection chamber. Prepare a single cell suspension and introduce a few hundred cells into the chamber and allow to settle onto the bottom surface.

2. To immobilize a single blastocyst, position onto the end of the holding pipet and move to the center of the field. Under high-power magnification, select individual cells carefully; suck sufficient cells for a single injection (between 1 and 15, depending on the experiment) into the injection pipet and position near the tip. Then align the injection pipet in the vertical plane by touching the end gently to the surface of the embryo to cause a small indentation, taking care not to damage the zona pellucida. This procedure gives a good indication of the position of the end of the pipet relative to the surface of the blastocyst. In this way the point can be moved into an "equatorial" position on a "trial and error" basis.

3. Gently introduce the injection pipet into the blastocoel and slowly expel the cells. Following withdrawal of the injection pipet, the blastocyst collapses and this results in the cells being brought into contact with the surface of the inner cell mass (ICM).

The complete injection procedure is illustrated in Figure 67. It is often convenient to introduce the pipet at a junction between two trophoblast cells, as damage to the embryo is minimized and successful penetration of the needle is easier.

Following this procedure it is routinely possible to inject expanded blastocysts at a rate of 30–40 per hour. Occasionally injection may be hampered by the accumulation of cellular debris around the end of the injection pipet. This results in a blunting effect and if the debris cannot be removed, either by purging the instrument with oil or by touching the needle end to the oil–medium interface, the best solution is to replace the injection pipet.

Injected embryos are removed periodically and cultured in drops of DMEM + 10% fetal calf serum at 37°C under oil. Following 1–3 hours in culture the blastocysts reexpand, and in many instances the injected cells can be seen adhering to the inner cell mass.

The injection procedure achieves best results when well-expanded blastocysts are used. The use of partially expanded blastocysts in the injection procedure may result in damage to the inner cell mass. In addition, when reexpansion occurs the trophoblast cells often exclude the introduced cells from the blastocoel. To ensure a supply of expanded blastocysts, embryos are collected 3.5 days p.c. For convenience, and to also allow blastocyst injections to begin in the morning, the mice can be maintained on a shifted 12-hour light-dark cycle so that the dark period ranges from 3 pm to 3 am. The majority of embryos will be fully expanded if the females are flushed at 9 am on the fourth day of pregnancy.

Figure 67   Procedure for injection of blastocysts. (*A*) A blastocyst is immobilized on the holding pipet such that the inner cell mass is positioned on the far left-hand side of the embryo. The injection needle is used to collect individual cells which are positioned near the needle tip. The tip of the needle is brought into the same focal plane as the midpoint, or equator, of the blastocyst. (*B*) With a single and swift and continuous movement the needle is introduced into the blastocoel cavity. (*C*) The cells (arrowed) are released slowly into the blastocoel cavity.

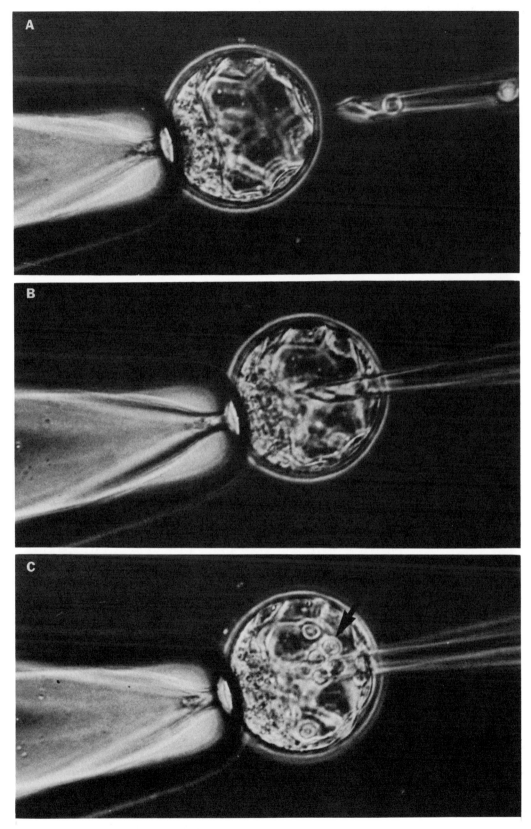

Figure 67   (*See facing page for legend.*)

Transfer of Operated Embryos into the Uterus

(see Section C, Uterine Transfer)

Following reexpansion, the operated embryos are surgically transferred into the uterine horns of pseudopregnant recipient females 2.5 days p.c. (Section C). In well-coordinated experiments, operated embryos are divided into groups of five to six and each group is returned to a single uterine horn of a recipient female. The minimum number of embryos required to maintain pregnancy is four in a unilateral transfer; the maximum number used is 12 embryos, transferred bilaterally. A suitable strain of mouse to use for recipients is (C57BL × CBA)F_1. Operated females are inspected after 14 days and pregnant animals are selected and caged individually. The pups are normally born 17 days posttransfer. The overall recovery rate for injected embryos is in the region of 50%, with a recipient failure rate of approximately 20% and an implantation rate of approximately 65% accounting for the embryonic losses.

NUCLEAR TRANSPLANTATION IN THE MOUSE EMBRYO

In this method* (McGrath and Solter 1983b) the pronuclei are removed without penetrating the plasma membrane of the egg. Instead, they are withdrawn into a membrane-bound karyoplast which can then be fused with a recipient enucleated egg using inactivated Sendai virus. This method depends critically upon preincubating the eggs in the presence of the cytoskeletal inhibitors, cytochalasin B and colcemid. Cell fusion induced by Sendai virus injected within the zona pellucida was first described by Lin et al. (1973).

EQUIPMENT

Soft flint glass tubing (1.0 mm o.d. and 0.6 mm i.d.) (from Drummond Scientific Company, Broomall, Pennsylvania 19008) is used for both holding and enucleation/injection pipets

Grinding wheel (Narishige EG-3)

DeFonbrune Microfuge with both thick (0.12–0.3 mm) and thin (0.1 mm) filaments and a supply of platinum wire for replacing filaments (see Section H, Supplies) (0.1 mm = 36-gauge platinum wire; 0.3 mm = 28-gauge)

Diamond paste (Astro-Met diamond compound; GCA/Precision Scientific, Chicago, Illinois)

HF acid, NP-40 detergent, distilled water

Beaudouin syringe (DeFonbrune)

Embryo culture medium (e.g., bicarbonate-buffered Whitten's medium [Whitten 1971] supplemented with 100 μM EDTA [Abramczuk et al. 1977])

Culture medium for microsurgery (HEPES-buffered Whitten's medium supplemented with cytochalasin B [5 μg/ml] and colcemid 0.1 [μg/ml] (McGrath and Solter 1983b)

*Information provided by Dr. James McGrath and Davor Solter, Wistar Institute, Philadelphia Pennsylvania.

Embryo Isolation

1. Isolate one-cell embryos and remove the cumulus cells, as described in Section C.

2. Culture the embryos under silicone oil (Dow Corning 200 Fluid 50 cs of viscosity) at $37\,^\circ$C in an atmosphere of 5% O_2, 5% CO_2, 90% N_2.

3. Prior to microsurgery the embryos are incubated at $37\,^\circ$C for 15–45 minutes in bicarbonate-buffered Whitten's medium containing 5 μg of cytochalasin B/ml and 0.1 μg Colcemid/ml.

4. Throughout the microsurgical manipulations, fix the embryos in place using a holding pipet, while either removing or introducing the nuclei via an enucleation/injection pipet. The microscope/micromanipulator set-up described for pronuclear injection can be adapted for nuclear transfer, with suction on the enucleation pipet being controlled by a Beaudouin syringe (DeFonbrune).

Making the Holding Pipet

1. Hand-pull capillary tubing over a microburner and break on a glass anvil to give an outside diameter of 75 μm using a DeFonbrune microforge (see Fig. 65).

2. Then polish the pipet tip on the microforge using a heated thick filament (0.3-mm platinum wire). (This holding pipet is essentially identical to the holding pipet described for the microinjection of DNA into pronuclei.)

Making the Enucleation/Injection Pipet

1. Pull capillary tubing on a vertical pipet puller (e.g., David Kopf Instruments, Model 700/D). Then break the tip of the pipet on a glass anvil using a DeFonbrune microforge (thin-filament, 0.1-mm platinum wire) to give an outside diameter of 10–15 μm at the tip. This process is diagramed in Figure 65.

2. Then bevel the pipette tip on a grinding wheel coated with a thin layer of diamond paste. Rinse the bevelled pipet tip (both inside and outside) in hydrofluoric acid (HF) to remove residual diamond paste and then rinse several times in distilled water. (*Note:* The concentration of HF acid used depends on relative hardness of the glass. For soft flint glass we typically use a 10% solution [vol/vol] of HF acid, whereas for Pyrex glass we use a 25% HF acid solution.) Visually inspect the pipet tip (the microforge can be used for this) to determine if the bevel and tip diameter are acceptable. "Thin" the pipet tip by repeated rinsing in HF acid with air blowing through. This means that only the outside is thinned. In order for the pipet to penetrate the zona, the tip of the micropipet must then be sharpened by touching it to the thin (0.1 mm) microforge filament. This process is diagramed in Figure 68.

Figure 68 Sharpening the enucleation/injection pipet. Position the beveled pipet vertically above the thin filament. Heat the filament to a temperature at which the pipet tip will just barely fuse to the filament (to be determined empirically). The air jets of the microforge must be blowing on the filament so that it heats the pipet only by conduction and not by convection. When the pipet fuses to the filament, raise it away from the filament, pulling out a short spike of glass at the tip.

3. After sharpening, rinse pipettes in 100% NP-40 detergent, rinse repeatedly in distilled H_2O, wipe with a Kimwipe, and store overnight. This procedure coats the surface of the pipet so that the egg membrane does not stick; the NP-40 treatment is used in place of a siliconizing agent.

Method for Enucleation

Pronuclei are removed in a manner that does not require penetration of the ovum plasma membrane.

1. Hold the embryo in position with the holding pipet and penetrate its zona pellucida with the enucleation/injection pipet. This may require considerable deformation of the zona (Fig. 69A–C). Avoid penetrating the embryo plasma membrane. The presence of the cytoskeletal inhibitors assists in the "nonpenetration" of the plasma membrane since in their presence the embryo offers little resistance to the advancing micropipet.

2. Once the pipet tip is in the perivitelline space (PVS), further advance it to a point adjacent to one of the two pronuclei. Then sequentially draw up into the pipet the plasma membrane overlying the pronucleus, a small volume of cytoplasm, and the pronucleus itself (Fig. 69D); use a Beaudouin syringe (De-Fonbrune) to control the suction. Then move the pipet tip to a point adjacent to the remaining pronucleus and repeat this step (Fig. 69E).

3. Withdraw the pipet from the embryo (Fig. 69F). As withdrawal proceeds a cytoplasmic bridge will be seen extending from the karyoplast within the pipet and the enucleated embryo within the zona pellucida. This bridge will stretch to a fine thread, pinch off, and reseal. This completes the enucleation step.

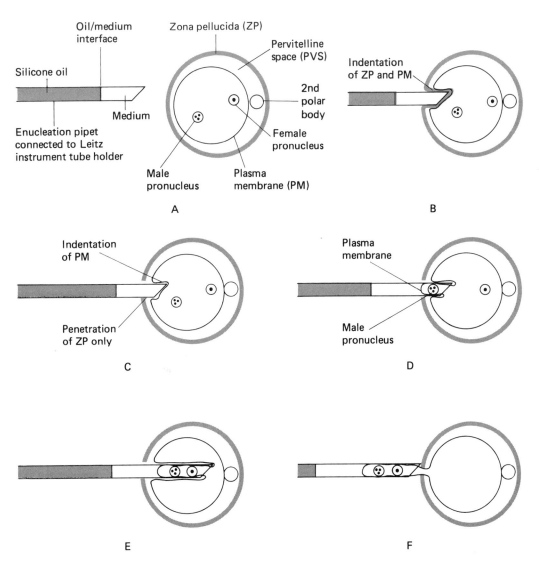

Figure 69 Technique for enucleating an egg by withdrawing pronuclei into a karyoplast.

Method for Introduction of Nuclei into Enucleated Embryos

Figure 70A–F illustrates how the pronuclear karyoplast obtained above can be introduced into a second enucleated one-cell-stage embryo. However, similar manipulations with nuclei obtained from other sources are also possible.

1. Move the pipet containing the pronuclear karyoplast to a drop containing inactivated Sendai virus (approximately 3000 HAU/ml) (Giles and Ruddle 1973). Draw a small volume of virus (approximately equal to the volume of the karyoplast) into the pipet (Fig. 70A,B).

2. Move the pipet to a third drop containing a second enucleated embryo. Position the embryo with a holding pipet so that the previous site of penetration is accessible to the pipet containing the karyoplast and virus. Advance the pipet through the hole in the zona pellucida, and sequentially inject the virus and karyoplast into the perivitelline space (Fig. 70C,D).

3. Then release the embryo from the holding pipet and transfer to culture medium (37°C) lacking cytoskeletal inhibitors. Visually inspect embryos on a dissecting microscope 1 hour later to determine the proportion of embryos that successfully incorporated the donor pronuclei (Fig. 70E,F).

Preparation of Inactivated Sendai Virus

Sendai virus is prepared as described by Giles and Ruddle (1973) and Graham (1971).

1. Inactivate virus (Neff and Enders 1968) by adding 1% β-propiolactone (BPL) (diluted in cold phosphate-buffered saline) to the virus suspension, on ice, to achieve a final concentration of 0.025% BPL.

2. Store virus-BPL mixture for 24 hours at 4°C, and then place it in a 37°C water bath for 20 minutes.

3. Subsequently aliquot inactivated virus (100 μl), fast-freeze in an acetone-dry ice bath, and store at −70°C.

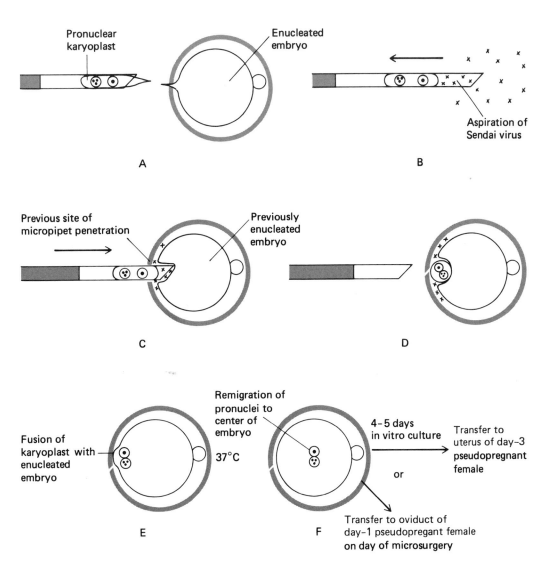

Figure 70   Technique for introducing nuclei from a karyoplast into an enucleated one-cell embryo by inactivated Sendai virus-mediated fusion.

Section E

ISOLATION OF PLURIPOTENTIAL STEM CELL LINES

Techniques have recently been described that make possible the isolation in culture of pluripotential stem cell lines. The technique described here for isolating such cell lines is essentially that originally reported by Evans and Kaufman (1981). Alternative methods, employing somewhat different experimental techniques, have also been reported (Martin 1981; Axelrod and Lader 1983). The stem cells can be injected into normal expanded blastocysts. They become integrated into the inner cell mass and subsequently contribute to many different tissues of the embryo, including the germ line.

ISOLATION OF STEM CELL LINES

The embryonic starting material can either be normal 3.5-day p.c. expanded blastocysts or "delayed" blastocysts (see later). Delayed blastocysts are normally collected 4–6 days following ovariectomy. For both groups the culture procedures are identical, with the only difference being the timing of the first disaggregation, as delayed blastocysts will initially grow more slowly.

All tissue culture takes place in a 37°C, 10% CO_2 humidified incubator and the culture medium used throughout is Dulbecco's modified Eagle's medium (DMEM) supplemented as described in the preceding section.

MATERIALS

1. Culture medium. The medium used throughout the culture procedure is Dulbecco's modified Eagle's medium (DMEM) (high-glucose, low-pyruvate formulation) supplemented with 10% fetal bovine serum, 10% newborn calf serum, 1× antibiotics, 10^{-4} M 2-mercaptoethanol, and 1% of a 100× stock of nonessential amino acids (Gibco).

   It is important to control the quality of the serum used for embryo culture, and batches of both fetal and newborn calf serum should only be purchased following rigorous testing. The most convenient way to compare different sera is to perform a comparative plating efficiency test using a stem cell culture that is seeded onto preformed feeder layers at clonal densities.

2. Trypsin. A trypsin/EDTA solution is used to dissociate cells. A final concentration of 0.25% trypsin, 0.4% EDTA in Tris-buffered saline is normally used. (Note that these concentrations are higher than those used in subculture of teratocarcinoma cells, Section G.)

3. Preparation of feeder layers. Feeder plates contain a layer of mitotically inactive fibroblast cells and appear to be necessary to maintain stem cells in exponential growth and to inhibit differentiation. Confluent plates of STO mouse fibroblasts are treated with freshly prepared DMEM + 10% newborn calf serum containing 10 μg/ml of mitomycin C (Sigma). The plates are returned to the incubator and left for 2–3 hours. The plates are then washed extensively with several changes of phosphate-buffered saline (PBS) and the cells collected by trypsinizing. The cells are sedimented by low-speed centrifugation (1000 rpm for 5 min), resuspended in fresh medium, counted, and diluted to give a final density of 3×10^5/ml. The cells are then plated out onto tissue culture dishes that have been pretreated with gelatin by flooding with a 0.1% solution of gelatin (Swine Skin type II, Sigma) for 2 hours. It is important to plate the fibroblasts at the correct density to ensure that a confluent uniform monolayer is produced. A density of 5×10^4/cm² is suggested as being suitable. Feeder plates

[handwritten margin note: ATCC Catalogue p 145]

Information supplied by Dr. Liz Robertson, Department of Genetics, University of Cambridge, Cambridge, U.K.

may be used up to 10 days after they are made and the medium should be replaced immediately before use. STO cells are a thioguanine- and ouabain-resistant subline of SIM mouse fibroblasts isolated by Dr. A. Bernstein (for references, see Martin and Evans 1975).

PROCEDURE

1. Flush embryos from the uterine horns (Section C) and place individually into 10-mm wells (Nunc 4×10-mm well plates are ideal) containing a preformed feeder layer and 1 ml of medium. The first stage of embryo culture can also be carried out in drops of medium incubated under light-weight paraffin oil. After 2 days of culture the embryos hatch from the zona pellucida and attach to the tissue culture surface by migration of the trophoblast cells. Shortly following attachment the inner cell mass (ICM) component becomes readily distinguishable and can be seen to grow rapidly over the following 2 days. Normally after this time the ICM component is considerably enlarged. However, embryos within a group may vary considerably, so that daily monitoring of individual embryos is essential. Examples of the progressive change in the morphology of cultured blastocysts are given in Figure 71A–D.

2. When the ICM-derived clump has attained the stage illustrated in Figure 71, C or D (a stage normally reached 4–5 days following explantation into culture), dislodge it from the underlying sheet of trophoblast cells using the sealed end of a finely drawn out Pasteur pipet. Wash the clump through two changes of Ca^{++}/Mg^{++}-free PBS and transfer to a small drop of trypsin solution under oil.

3. Incubate 3–4 minutes at 37°C, and then use a finely drawn out Pasteur pipet (diameter of end > cell clump), prefilled with serum-containing medium, to disaggregate gently the clump into smaller cellular aggregates of three to four cells. It is not advisable to attempt to reduce the ICM clump to a single cell suspension.

4. Transfer the contents of the drop to a fresh 10-mm feeder well. Inspect the individual cultures daily.

Generally, after 2 days primary colonies of cells will become readily visible. These can have one of several different morphologies: (1) Trophoblast-like cells; (2) epithelial-like cells; (3) endoderm-like cells; and (4) stem cell-like cells.

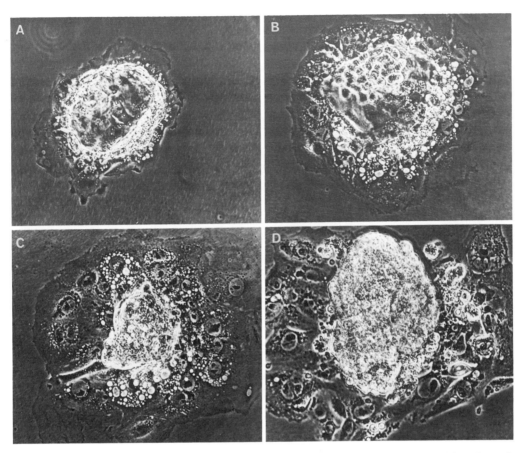

Figure 71 Progressive changes in the morphology of cultured blastocysts. (*A*) 48-hr culture; (*B*) 72-hr culture; (*C*) 96-hr culture; (*D*) 120-hr culture. When the inner cell mass-derived component has attained approximately the size and morphology illustrated in *D*, it is suitable for removal and disaggregation as described in the text.

Trophoblast-like Cells

In almost 100% of cases areas of trophoblast-like cells rapidly become apparent. These cells are morphologically identical to those seen in the initial blastocyst outgrowths. Examples of the morphology of these trophoblast giant cells are given in Figure 72. Frequently presumptive stem cell colonies can be located during the early stages of culture. The cells in these colonies initially appear to be identical in morphology to pluripotential stem cells (Fig. 72). However, if such colonies are closely observed, within 1–2 days the cells at the periphery of the colony will flatten and spread; generally within the following 24 hours the colony rapidly differentiates to form giant cells exclusively.

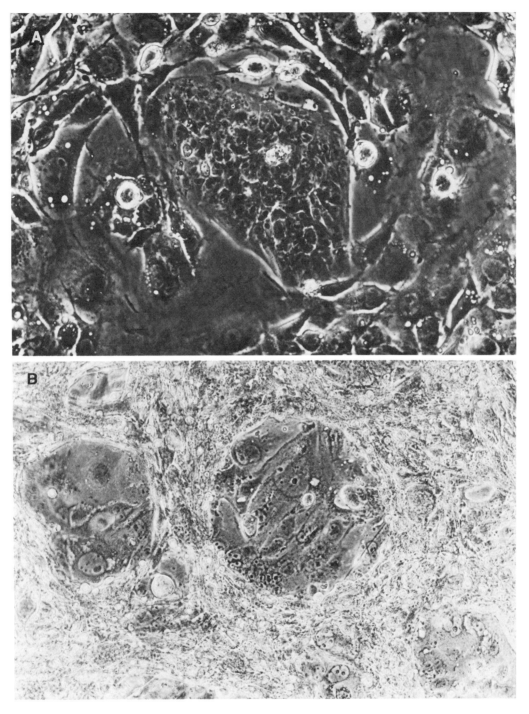

Figure 72 Morphology of primary cell colonies obtained following dissociation of the inner cell mass. (*A, B,* and *C*) Examples of "trophoblast-like" colonies. These colonies may first be apparent 1–2 days after passage as small nests of cell that have a superficially "stem-like-cell" morphology (*A*). With further culture the cells alter in appearance and flatten and spread to give areas composed of a monolayer of giant cells (*B,C*). Although embryo-derived cells of this phenotype grow well in initial cultures, they fail to proliferate with long-term culture. (*D*) "Epithelial-like" cells. This cell type forms discrete colonies that grow comparatively slowly. If cultured for long periods (2 weeks), large, flat colonies composed of a monolayer of cells will result. (*See next page for* C *and* D.)

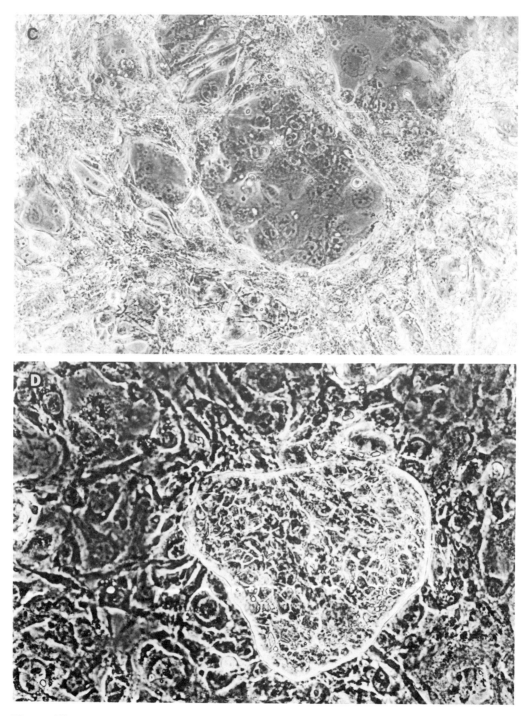

Figure 72 (Continued)   (See previous page for legend.)

Epithelial-like Cells

Occasionally colonies of a very distinct and easily recognizable phenotype will form. These cells are very slow growing and form discrete patches on the feeder cell layer. The constituent cells pack together to give a flat pavement, epithelial-like structure, often with a very marked, highly refractile edge to the colony. An example of a typical colony is given in Figure 71D.

Endoderm-like Cells

Areas of rounded, refractile, loosely attached cells grow in a few cultures. These appear to be similar to the endodermal cell type that forms when stem cells are encouraged to differentiate in culture.

Stem Cell-like Cells

These are cells that grow progressively but maintain a stable stem cell, or embryonal carcinoma (EC) cell, phenotype. Such colonies are illustrated in Figure 73. Discrete colonies of a stem cell morphology are located and marked and those that fail to show any differentiation but retain cells of an exclusively EC phenotype are selectively removed after a total of 7–8 days of culture. These colonies are then dissociated in drops of trypsin, as described above, and passaged into fresh feeder wells. In successful cultures small nests of stem cells appear within 2–3 days of subculture. Depending on the relative rate of growth these cultures are expanded 3–5 days later by trypsinizing the whole well and transferring the contents onto a 3-cm feeder plate. The experimental stages of the technique and the timings of the various disaggregation events are summarized diagramatically in Figure 74.

An established stem cell line requires careful subculture at 3- to 6-day intervals, depending on the growth rate. If care is not taken on passaging, the cells may differentiate extensively to form large quantities of endoderm. To improve the viability of the cells, confluent cultures are generally refed 2–3 hours prior to passage. The cell monolayer is washed well with two changes of PBS and a small quantity of trypsin is added to the dish. After incubation at 37 °C for 3–4 minutes, the cells can be seen to detach from the plate. The cell clumps are broken up to give a single cell suspension by the addition of a small quantity of serum-containing medium and pipeting the suspension vigorously using a Pasteur pipet. The cell suspension is reseeded onto fresh feeder plates. Cultures are generally refed at 2- to 3-day intervals.

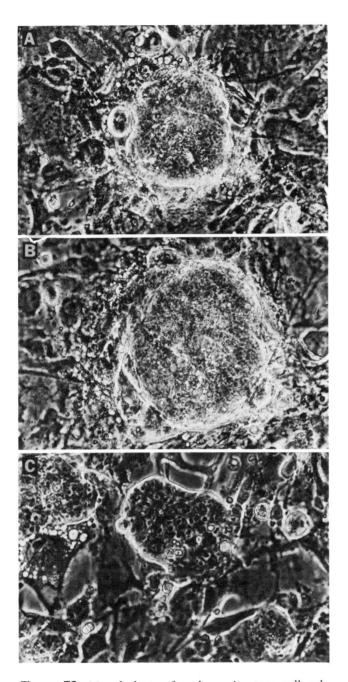

Figure 73 Morphology of embryonic stem cell colonies. (*A*) Appearance of a colony of stem cells 2 days after the disaggregation of the inner cell mass. (*B*) Same colony 2 days later. Note that the colony remains composed of a homogeneous population of stem cells and that no overt cellular differentiation has occurred. Stem cells are comparatively small, typically have a large clear nucleus containing one or more prominent nucleoli, and are tightly packed within the multilayered primary colony. (*C*) The colony shown above in B was subcultured (as described in the text) into a fresh feeder well. Within 2 days numerous small nests of stem cells (illustrated) appear in the culture.

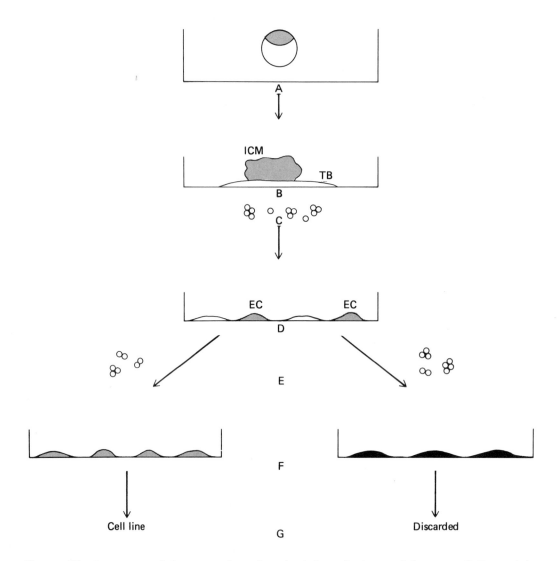

Figure 74 Summary of the procedure for obtaining pluripotential stem cell lines. (*A*) Day 0, blastocysts recovered and placed into feeder wells. (*B*) Day 5–6, ICM has proliferated to give a large cellular mass. (*C*) Day 5–6, ICM clump is removed from trophoblast, outgrowth dissociated, and cellular fragments transferred to fresh feeder well. (*D*) Day 9, primary colonies examined and classified according to morphology. (*E*) Day 12, colonies that resemble stem cell (EC) are individually removed, dissociated, and reseeded into fresh feeder wells. (*F*) Day 14, wells inspected for the presence of stem cells. (*G*) Day 16, wells containing stem cells are subcultured to give permanent cell lines. Remaining wells are discarded.

OVARIECTOMY OF PREGNANT FEMALES TO INDUCE IMPLANTATIONAL DELAY OF BLASTOCYSTS

The surgical removal of ovarian tissues, in combination with the administration of progesterone (Depo-Provera), causes artifical arrest of blastocyst development. The viability of the embryos is maintained for up to 10 days following ovariectomy. This technique is described more fully by S. Bergstrom (1978).

EQUIPMENT

Avertin
Animal balance
1-cc syringe
26-gauge 1.3-cm needle
70% ethanol
Blunt-ended forceps
Iris scissors
Two pairs of #5 watchmaker's forceps
Two loops of surgical suture silk
Surgical needle and silk
Wound closing clips
Depo-Provera solution of 10 mg/ml in sterile saline (Upjohn Labs)

This procedure is easily performed without the need for magnification, but a low-power binocular dissecting microscope may be useful.

PROCEDURE

The operation is carried out on the afternoon of the third day of pregnancy (i.e., 2.5 days p.c.). This ensures that the morulae have traveled to the end of the oviducts and will be less likely to be damaged during the ovariectomy procedure.

1. Anesthetize the mouse with a standard dose of Avertin (see Section C, Embryo Transfers).

2. Wipe the back with 70% alcohol and make a small lateral incision through the skin of the back. Sliding the skin from side to side, locate the position of the ovaries beneath the peritoneal wall. Make a small incision over one ovarian fat pad, grasp the fat with the blunt forceps, and pull out the ovary and oviduct.

3. Next, tear the membrane surrounding the ovary and oviducts using two pairs of fine forceps. This releases the ovary. The intact ovary and ovarian blood vessel are tied by slipping the loop of suture silk between the oviduct and ovary. Cut ovarian tissue away using fine scissors.

4. Replace the uterine horn in the abdomen using the blunt forceps and close the peritoneal incision with a single suture.

5. Repeat the procedure for the other side and close the incision in the back with a wound clip.

6. While the female is still anesthetized, inject 1 mg of Depo-Provera in 0.1 ml sterile saline subcutaneously into the flank region.

Summary

Schedule for collection of delayed blastocysts

Day 0 Female caged with fertile male

Day 0.5 Plug checked

Day 2.5 Ovariectomy and administration of 1 mg of Depo-Provera

Day 6.5–10.5 Delayed blastocysts recovered by flushing uterine horns

Section F

TECHNIQUES FOR VISUALIZING GENES AND GENE PRODUCTS

Much valuable information on the specificity of expression of endogenous and foreign genes can be obtained by analysis of RNA extracted from tissues at different stages of development. However, the amount of material available from mouse embryos is usually very small, and the gene may be active in only a minority of the cell population. To overcome these problems it is necessary to turn to in situ hybridization of embryo sections and the following pages provide the simplest of guides to this technically demanding and specialized field. Another important use of in situ hybridization is to map cloned genes to mouse chromosomes, and to localize the integration site of foreign DNA in transgenic mice. Finally, we briefly outline methods for visualizing antigens in tissue sections using immunofluorescence light microscopy.

THE KARYOTYPING OF MOUSE CELLS

The normal mitotic karyotype consists of 40 acrocentric chromosomes which, in an average preparation, range in length from 2 to 5 μm. Since after conventional staining they show few distinguishing morphological features (e.g., see Evans 1981), a complete analysis requires chromosome banding, and of the numerous available methods (e.g., see Miller and Miller 1981) Q and G banding have proved the most popular. The latter gives the greater resolution and has therefore provided the foundation for the standard idiogram of banding patterns. A list of 312 distinct regions defined by G banding within the karyotype has most recently been published, together with the rules for the nomenclature of chromosome anomalies, in Green (1981). As yet, apart from the consistent strain variations observed in the centromeric regions of some homologs, no other variations have been detected in the standard, normal chromosomes within and between strains of the laboratory mouse (e.g., see Evans 1981) and any differences observed should be regarded as either an abnormality or a sign of karyotypic evolution (Fig. 75).

Providing there is mitotic activity, chromosome preparations can be made from all tissues of the mouse whether they are derived from in vivo or in vitro sources. In general, the greater the mitotic activity the more likely the success but, unfortunately, the mitotic metaphase arrestants such as colcemid, colchicine, and vinblastine sulfate, which can greatly enhance the frequency of desired cells, also have the concomitant and deleterious effect of contracting the chromosomes, thus reducing subsequent G-band resolution. If possible, the use of mitotic arrestants should be avoided or, if necessary, be limited to a brief exposure in both time and concentration.

The majority of mouse chromosome preparations for banding are now made by air-drying and in essence require the production of a cell suspension as a starting point. Some samples such as blood cultures, ascitic fluids, or cells growing in suspension will already be in this state; others, such as bone marrow, solid tumors, or cells growing as attached layers, will have to be converted into suspensions. This can be achieved by using a variety of disaggregation methods, some of which are, for example, described by Cronmiller and Mintz (1978). The basic steps in karyotyping embryonal carcinoma cells are outlined below.* Innumerable variations of these essential technical steps have been published, for example by Cronmiller and Mintz (1978), McBurney and Rogers (1982), and Robertson et al. (1983).

*Information provided by Dr. E.P. Evans, Sir William Dunn School of Pathology, University of Oxford, Oxford OX1 3 RE, U.K.

X 1,300

X 2,600

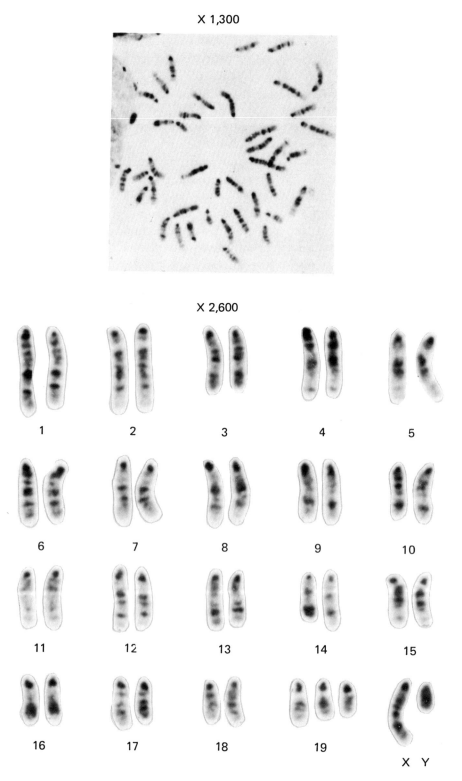

1   2   3   4   5

6   7   8   9   10

11   12   13   14   15

16   17   18   19

X Y

Figure 75 G-banded cell from an embryonal carcinoma cell line which, when first derived, had 40 normal chromosomes and was XY. After continued culture the karyotype, although remaining normal as far as the G bands were concerned, evolved by acquiring an extra copy (trisomy) of chromosome 19.

EQUIPMENT

Acid-washed, grease-free glass slides (for preparation, see below)
Paper tissues (grease-free)
Pasteur pipets
Desk lamp
Microscope with 100× oil immersion objective and phase-contrast
  objective (40×)
Coplin staining jars

SOLUTIONS

0.56% (wt/vol) KCl (hypotonic solution for swelling cells)
3:1 Mixture of absolute methanol and glacial acetic acid (fix)
2× SSC (0.3 M NaCl, 0.03 M trisodium citrate)
0.85% (wt/vol) NaCl
0.025% Trypsin (Difco 1:250) in 0.85% NaCl (the age of the trypsin
  powder is not important, but the solution should be made up
  approximately 30 min before use)
1:30 Giemsa stain (Merck) in 5 mM phosphate buffer (pH 6.8) (the
  buffer can conveniently be obtained in tablet form from Gurr)

PROCEDURE

1. Scrape off in their growth medium embryonal carcinoma cells, growing in tissue culture dishes and without pretreatment with a mitoic arrestant. Centrifuge at 1000 rpm for 5 minutes (a speed and time used throughout the procedure) in a 2-ml conical, glass tube.

2. Resuspend cell pellet in aqueous 0.56% (wt/vol) potassium chloride as the hypotonic solution and leave at room temperature (21–23 °C) for 6 minutes.

3. Centrifuge again, carefully remove the supernatant, and fix the cell pellet in a freshly made mixture of three parts of absolute methanol to one part of glacial acetic acid at room temperature. For the fixation procedure, gently pipet in the fixative without disturbing the pellet and replace with two further changes of fixative over a period of 3 minutes. After this time, remove the fixative and resuspend the "dry" pellet by vigorously flicking the tube with the forefinger while simultaneously adding more fixative.

4. Finally, centrifuge the cells and resuspend in a small volume (0.5 ml) of fixative.

   For slide making:

1. Clean commercially purchased precleaned slides further by leaving overnight in a mixture of one part of absolute alcohol to one part of concentrated hydrochloric acid, wash the next day in running tap water, rinse in deionized water, and store in a mixture of one part of absolute alchohol to one part of diethyl ether.

2. Shortly before using, remove slides from the mixture with forceps and wipe dry on Kimwipe paper tissues (Kimberly-Clark), or some other coarse, grease-free tissues, avoiding finger contact to prevent the possibility of grease being applied to the surfaces.

3. To make chromosome spreads on these slides, add a row of three drops of cell suspension from a small glass pipet (approximately 0.2 ml capacity and drawn to deliver drops of 10 μl), letting them spread to their maximum and begin to dry until interference rings are visible at their periphery. Assist the drying by blowing from the mouth while holding the slide up to the heat generated by a desk lamp bulb.

4. When dry, inspect the slides under low-power phase-contrast microscopy (final magnification 160×) for cell quality and quantity. If necessary, cell density can be increased by adding and drying further drops of the cell suspension.

   Preparations are best made within 0–3 hours of the final fixation, but can be made up to 7 days afterwards providing the material is kept at 4 °C and the cell pellet resuspended in freshly made fixative before attempting to spread the cells. The longer the material is kept, the poorer the spreading quality of the mitotic cells. If possible, at least five slides should be made from each sample to guarantee success. Before G-banding, slides should be "aged" for between 3 and 21 days by leaving in a closed box at room temperature. Fresh slides give poor G-band resolution, maximum resolution is achieved at approximately 10 days and afterwards slowly declines until after several weeks in storage the chromosomes either fail to band and stain uniformly or show "pseudo-bands" that bear no meaningful relationship to the standard idiogram.

For G banding:

1. Place suitably aged slides in $2\times$ SSC (0.3 M sodium chloride, 0.03 M trisodium citrate) in a lidded coplin jar in a water bath at 60–65 °C for 1.5 hours. Then cool to room temperature by running tap water over the closed jar and transfer to 0.85% (wt/vol) NaCl for 5 minutes at room temperature.

2. Drain slides by touching onto filter paper, place in a horizontal plane, and flood with 0.025% trypsin solution for between 15 and 20 seconds. The trypsin time is critical, since underexposure preserves chromosome morphology but gives poorly differentiated bands, whereas overexposure distorts morphology and eliminates most of the bands. Optimum trypsin times are known to vary between laboratories and a test slide should be treated for the minimal suggested time of 15 seconds to establish the best treatment time for the rest of the slides. In the laboratory of Dr. E.P. Evans, the optimal time has been established as between 15 and 20 seconds for mouse chromosomes, and has remained constant, irrespective of the source of the mitotic cells, for the last decade.

3. Stop tryptic activity by placing the slides back into normal saline and then rinse slides in 6.8 pH phosphate buffer and stain in freshly made Gimesa (Merck) at a dilution of one part of stain concentrate to 30 parts of 6.8 pH buffer. After 10 minutes in the stain, monitor wet slides under low-power, bright-field microscopy ($160\times$) for staining intensity. Since, on drying, wet slides gain contrast, care should be taken not to overstain as this will reduce G-band differentiation. If necessary, repeat staining until adequate results are achieved and then quickly rinse in 6.8 pH buffer and blow dry with a current of cool air.

4. Examine unmounted slides under a $100\times$ oil immersion lens or mounted in a Xylol-based mounting medium such as Eukitt (Reidel-De Haen Ag Seelze-Hannover). The majority of modern, readily available immersion oils which are declared "PCB free" also have the unfortunate property of removing Giemsa stain after a few hours' exposure. Although direct viewing in immersion oil gives a higher optical resolution, it is wiser to mount slides if they are to be kept.

5. For photographing suitably spread and G-banded mitotic cells, use a green Balzer interference filter (B40 5268) and Kodak technical pan film 2415 which is developed in Kodak HC110 developer at the recommended high-contrast dilution, D.

MAPPING INTEGRATED GENES TO CHROMOSOMES BY IN SITU HYBRIDIZATION

Before hybridization, G-banded metaphase spreads are photographed and their position on the slide recorded with the aid of graticule.

RNasing the Slides

1. Destain slides: $3\times$ 5-minute washes in 70% EtOH; $1\times$ 5-minute wash in 95% EtOH. Air dry.

2. Immerse slides in 250 ml of $2\times$ SSC containing 100 μg/ml of RNase A. Incubate for 1 hour at 37°C or 90 minutes at room temperature. (Preboil the stock RNase, 5 mg/ml in H_2O, to get rid of the DNase.)

3. Rinse: $10\times$ with $2\times$ SSC (a few seconds for each); $3\times$ with 70% EtOH (5 min each); $2\times$ with 95% EtOH (5 min each). Air-dry and put back into boxes and dessicator until ready to denature and hybridize.

Hybridization

For denaturation:

1. Bring 285 ml of formamide to pH 6.5 with drops of concentrated HCl. Allow ammonia fumes to escape; no odor should be present.

2. Add 15 ml of $20\times$ SSC alternately with drops of concentrated HCl to pH 7.0–7.2.

3. Denature slides by immersing them in the 95% formamide–$1\times$ SSC solution. Bring slowly to 70°C, i.e., start H_2O bath at room temperature and bring to 70°C. Leave at 70°C for 2–2.5 hours.

4. Immerse the slides in a dish of 250 ml of 70% EtOH that has been stored at -20°C. Allow to stand on ice for 10 minutes. Then wash: $2\times$ with 70% EtOH, 10 minutes each; $2\times$ with 95% EtOH, 5 minutes each. (Also store the EtOH for these washes at -20°C and perform the washes on ice.)

5. Air dry the slides.

For hybridization:

1. Denature the probe (double-stranded DNA labeled to a specific activity of 10^9 dpm/μg by nick-translation with [^{125}I]dCTP) by incubating it at room temperature for 15 minutes in 0.1 N NaOH. Add 1 M phosphate buffer to 0.18 M and 1 N HCl to approximately 0.08 N to neutralize. The final concentration of the probe should be 20–40 ng/ml in: 50% formamide, $5\times$ SSC, $1\times$ Denhardt's, 100 μg/ml denatured salmon sperm DNA, 0.25 mM iododeoxycytidine, 20 mM phosphate buffer, and 10% dextran sulfate.

2. Add 75 μl per slide of the hybridization mix and cover with a coverslip. Keep the neutralized probe on ice and just before placing the complete hybridization mix on a slide, add the appropriate volume of the probe to an aliquot of the other ingredients in the hybridization mix, e.g., 3 μl containing 1.5–3 ng of probe and 72 μl mix.

3. Place the slide in a petri dish on buffer (5× SSC, 50% formamide)-saturated 3MM paper. Then place this dish in a box containing 20–40 ml of buffer. Tape the box closed.

4. Incubate at 42°C for 12–18 hours.

Washes

1. Place slides in a dish containing 200 ml of 5× SSC–50% formamide and remove coverslips.

2. Suspend slides in a beaker containing 2 liters of 2× SSC. Wash for 2 hours at room temperature with stirring and with three buffer changes.

3. Do two 10- to 15-minute washes in 2× SSC at 68°C in 250-ml dishes.

4. Cool slides to room temperature in 2× SSC.

5. Wash 3× with 70% EtOH, 5 minutes each. Wash 2× with 95% EtOH, 5 minutes each.

6. Air-dry the slides.

Autoradiography

1. Dip slides in 0.1% gelatin–0.01% chromalum. Dissolve gelatin in boiling H_2O. When it has cooled, add chromalum (chromium potassium sulfate, $CrK(SO_4)_2 \cdot 12H_2O$). Dip slides for 1–2 seconds; drain dry.

2. Dip slides in emulsion. Use either Kodak NTB-2 or Ilford L4. L4 is more sensitive and gives smaller grains than NTB-2, but the background is worse. Expose at 4°C for 2–8 days.

For further information regarding this technique, see Gerhard et al. (1981), Robins et al. (1981), and Wahl et al. (1982).

LOCALIZATION OF GENE TRANSCRIPTS IN EMBRYO SECTIONS: IN SITU HYBRIDIZATION WITH RNA PROBES

This protocol* gives only the most basic outline of a procedure for in situ hybridization using single-stranded RNA probes on cryostat sections of fixed, frozen mouse embryos. The advantage of the technique described here is that it is relatively quick and simple; however, the histological preservation of the tissue is not ideal. Better resolution at the cellular level would be obtained by dehydrating the fixed embryos and embedding them in wax before sectioning; see, for example, Cox et al. (1984) and Ingham et al. (1985).

The method is based on the hybridization of single-stranded, antisense RNA probes to single-stranded RNA. Therefore, great care must be taken to avoid contaminating sections, slides, solutions, etc., with RNase. Gloves should be worn throughout and changed frequently, and disposable plasticware and pipets used wherever possible. All solutions should be sterilized by autoclaving or filtration through a 0.22-μm filter, and then protected from contamination with RNase or bacteria.

PROTOCOL OUTLINE

A. General equipment

B. Chemicals and solutions for fixation, sectioning, and section pretreatment

C. Embryo fixation and embedding

D. Sectioning and postfixation

E. Section pretreatment

F. Probes

G. Hybridization

H. Posthybridization washing

I. Autoradiography—general equipment and solutions

J. Autoradiography—dipping slides

K. Autoradiography—developing

L. Staining and microscopy

---

*Protocol provided by Peter Holland, Laboratory of Molecular Embryology, National Institute for Medical Research, London. Based on methods in Roger (1979); Hafen et al. (1983); Cox et al. (1984); Brûlet et al. (1985); Ingham et al. (1985).

A. GENERAL EQUIPMENT

1. Stainless steel slide racks for slides and coverslips

2. Plastic or glass staining jars (e.g., 120 × 120 × 60 mm). These should be very clean and rinsed well with sterile distilled water. If necessary, soak glass jars overnight in water containing 0.1% diethylpyrocarbonate (an RNase inhibitor) and then autoclave.

3. Clean, "subbed" glass slides. Standard glass slides (e.g., 76 × 26 mm) are cleaned by submerging them at least overnight in metal racks in chromium trioxide solution in 85% wt/vol H_2SO_4. Rinse in tap water for 2 hours. Wash 3 × 30 min in sterile distilled water (dH_2O) and then bake in an oven at 150°C for 60 min and store in a clean, dust-free box. The day before use, pretreat or "sub" slides with poly-L-lysine. Dip a rack of slides into 0.25 M NH_4Ac, dry at 60°C in a dust-free box, and soak for 30 min in freshly prepared 50 μg/ml poly-L-lysine solution (see below) at room temperature. Air-dry overnight at room temperature and store in a dust-free place. Subbed slides should be used as soon as possible. The poly-L-lysine coating binds the tissue sections, which otherwise tend to float off the slides during the washes.

4. Glass coverslips (e.g., 22 × 22 mm). These should be cleaned in chromium trioxide and rinsed in tap water and distilled water, as above. Instead of being treated with poly-L-lysine, the coverslips are siliconized so that they can be removed without detaching the sections. After dipping in siliconizing solution (e.g., Repelcote), the coverslips are air-dried and baked (150°C for 2 hr).

5. Cryostat and knife, fine paint brush, diamond pen, several mounting blocks.

6 Plastic embedding molds.

7. Liquid N_2.

8. Waterbaths and ovens. Also incubator (50°C) and hot plate (50°C).

9. Sandwich boxes and silica gel desiccant.

B. CHEMICALS AND SOLUTIONS FOR FIXATION, SECTIONING, AND SECTION PRETREATMENT

1. 10× Phosphate-buffered saline (PBS) stock: 1.3 M NaCl, 70 mM Na_2HPO_4, 30 mM $NaH_2PO_4 \cdot 2H_2O$ made up in glass-distilled water. Autoclave and check that pH is 7.0.

2. 4% Paraformaldehyde (electron microscopy grade, e.g., TAAB Laboratories Equipment Ltd.). THIS MUST BE MADE UP FRESH BEFORE USE. Dissolve in 1× PBS (takes 2–3 hr at 60–80°C with agitation). Check that pH is 7.0. *Note:* The fumes are very toxic and the solution should be handled with care in a fume hood. Cool on ice before use.

3. Freshly prepared 0.5 M sucrose in phosphate-buffered saline (PBS).

4. Poly-L-lysine (Sigma Cat. No. P1274). Make a stock solution at 10 mg/ml in sterile water. Store frozen at −20°C. Before use dilute to 50 μg/ml in sterile 10 mM Tris pH 8.

5. 20× P buffer: 1 M Tris (pH 7.5), 0.1 M EDTA.

6. Pronase. Sigma type XIV protease (Cat. No. P5147). This should be predigested at 40 mg/ml in dH₂O for 4 hours at 37°C to destroy RNase, lyophilized, and stored dry in 40-mg aliquots at −20°C. Immediately before use, make up enough to fill a staining jar at 0.125 mg/ml in 1× P buffer (e.g., 40 mg in 320 ml). *Note*: It may be necessary to titrate the digestion of sections for optimal results; a concentration half that which completely removes the sections from the slides is about optimal.

7. 20× SSC: 3.0 M NaCl (175.3 g/liter), 0.3 M Na₃ citrate (88.2 g/liter).

8. 10% wt/vol glycine stock solution in distilled water.

9. 2 M triethanolamine-HCl (pH 8).

10. Acetic anhydride.

11. Tissue-Tek O.C.T. embedding medium for frozen tissue specimens (Miles Scientific).

12. Absolute ethanol.

13. Concentrated HCl.

C. EMBRYO FIXATION AND EMBEDDING

1. Dissect out the embryos and fix in fresh 4% paraformaldehyde in 1× PBS at 4°C for 18–24 hours. No comparative study has been made of fixation protocols for mouse embryos. It may be possible to shorten fixation time, e.g., to 20 minutes. Other fixatives could also be tried, e.g., 1% glutaraldehyde (Cox et al. 1984). For postimplantation stages up to about 9.5 days postcoitum (p.c.), it is easiest to fix and section the embryo within the deciduum. For orienting the embryo, the widest part of the pear-shaped deciduum is "top" (see Fig. 43).

2. Rinse in PBS and transfer to 0.5 M sucrose in 1× PBS at 4°C for 24 hours.

3. Blot off excess liquid and orient embryo or deciduum in a plastic mold filled with O.C.T.

4. Touch the mold onto the surface of liquid nitrogen and hold steady until frozen solid. Store at −70°C. Blocks frozen for 2 months have given satisfactory results.

D. SECTIONING AND POSTFIXATION

1. Remove embryo blocks from $-70\,^{\circ}$C to $-20\,^{\circ}$C, trim with a razor blade, mount onto cryostat chuck using O.C.T. and rapid cooling on dry ice, and trim further. Cut sections at $-20\,^{\circ}$C at approximately 8–10 μm. The initial sections through the deciduum can be cut quickly and collected onto unsubbed slides for monitoring progress into the block. When the required part of the embryo is finally reached, pick up individual sections by slowly bringing a subbed slide at room temperature next to the section. Several sections can be collected on one slide. It is convenient to mark the upper surface of the slide with a diamond pen.

2. Place slides on a hot plate at $50\,^{\circ}$C for 1–3 minutes for rapid drying, and then dry more slowly at room temperature in dust-free place for 1–2 hours. The slides should then be taken straight into postfixation and dehydration. They are handled either in metal racks in staining jars, or (if only a few) in 50-ml plastic disposable screw-capped tubes.

3. Immerse in fresh 4% paraformaldehyde in PBS for 20 minutes, and then once through the following: $3\times$ PBS (5 min), $1\times$ PBS (5 min), $1\times$ PBS (5 min), 30% EtOH (2 min), 60% EtOH (2 min), 80% EtOH (2 min), 95% EtOH (2 min), 100% EtOH (2 min).

4. Air-dry slides and store in a box containing desiccant at $-20\,^{\circ}$C for up to a few weeks, or use immediately.

E. SECTION PRETREATMENT

The purpose of the Pronase digestion is to remove protein associated with the mRNA that might inhibit hybridization. This step is one that can be varied to improve results, and in any case the length of incubation with enzyme may have to be changed for different tissues. The acetylation step is crucial for preventing the probe from sticking all over the slide, including the area outside the section. As in D, the slides can be handled in staining jars or in a 50-ml tube.

Method

Bring slides (still in desiccated box) to room temperature. Take through the following prehybridization steps, all at room temperature unless otherwise stated:

1. 0.2 M HCl (20 min) to remove basic proteins.

2. dH$_2$O (5 min).

3. 2× SSC (30 min at 70°C). The solution should be preheated to 70°C and the incubation carried out in a water bath.

4. dH$_2$O (5 min).

5. 0.125 mg/ml Pronase in 1× P buffer (10 min).

6. 0.2% glycine in 1× PBS (30 sec to block Pronase reaction).

7. 1× PBS (30 sec).

8. 1× PBS (30 sec).

9. 4% Paraformaldehyde in PBS (20 min).

10. 1× PBS (3 min).

11. Acetylate for 10 min in a *fresh* mix of acetic anhydride diluted 1:400 in 0.1 M triethanolamine (pH 8) (add the acetic anhydride to the aqueous triethanolamine *immediately* before use).

12. 1× PBS (5 min).

13. 30%, 60%, 80%, 95%, 100% EtOH (2 min each).

14. Air-dry (1–2 hr).

15. Use immediately for hybridization.

F. PROBES

For synthesis of single-stranded RNA probes, the clone should be inserted into a vector containing a promoter specific for SP6 RNA polymerase, for example, pSP64 and pSP65 (Promega Biotec; Melton et al. 1984). Alternative vectors are also available containing promoters for T7 and/or T3 polymerase (e.g., Gemini vectors, Promega Biotec; Bluescribe vectors, Vector Cloning Systems). The clone must be inserted in the orientation that allows synthesis of antisense probes, i.e., complementary to the mRNA to be detected. Control probes can be synthesized from clones inserted in the opposite orientation. Details of subcloning, bacterial transformation, and isolation of plasmid DNA can be found in Maniatis et al. (1982).

Solutions for Transcription Reaction

When ^{35}S-labeled probes are prepared, it is very important to make all solutions 10 mM with dithiothreitol (DTT).

1. Clone, digested to completion downstream of coding sequence. To avoid termination of the SP6 polymerase, it is very important that the DNA is clean and free of salt after the restriction digestion. Therefore, phenol extract, chloroform extract, ethanol precipitate, wash pellet in 70% ethanol, dry, and redissolve at 1 μg/μl in sterile dH$_2$O.

2. 1 M DTT (filter sterilize, and store at $-20\,^\circ$C in aliquots).

3. 10 mM CTP, ATP, GTP (e.g., P-L Biochemicals) in sterile dH$_2$O. Adjust pH to 7. Store at $-20\,^\circ$C in aliquots.

4. SP6 RNA polymerase.

5. RNase inhibitor (e.g., Amersham placental RNase inhibitor at 30 units/μl).

6. 5× Buffer: 200 mM Tris (pH 7.5), 30 mM MgCl$_2$, 10 mM spermidine-HCl, 0.1% Triton X-100.

7. RNase-free DNase I (Worthington) 1 mg/ml in sterile dH$_2$O. Store in aliquots at $-20\,^\circ$C.

8. Yeast total RNA. This is added to mop up any RNase contamination and as a carrier for ethanol precipitation. Remove any contaminating protein beforehand as follows: dissolve at 50 mg/ml in TE (pH 7.6) with 1% SDS. Digest with 100 μg/ml proteinase K for 3 hours at 37$\,^\circ$C, phenol extract, and ethanol precipitate. Dry and make up to 50 mg/ml in sterile H$_2$O. Store at $-20\,^\circ$C in aliquots.

9. 5 M NH$_4$Ac.

10. Absolute EtOH.

11. 10 mM Tris (pH 7.6), 1 mM EDTA (TE buffer).

12. Phenol/chloroform/isoamylalcohol (25:24:1) saturated with TE buffer.

13. 3 M NaAc (pH 5.2).

14. Uridine, 5'-[α-thio]-triphosphate, [^{35}S]—Specific activity 1000–1500 Ci/mmole.

15. Formamide, 2× deionized, 2× recrystallized.

Method

1. *Transcription*: For a total volume of approximately 20 μl. Make up at room temperature:

5× Buffer	4.0 μl
1 M DTT	0.2
RNase inhibitor (30 units/μl)	2.0
10 mM GTP	1.0
10 mM ATP	1.0
10 mM CTP	1.0
Restricted DNA (1 μg/μl)	1.0
[α-^{35}S]UTP	10.0
SP6 polymerase (16 units/μl)	1.0

Incubate at 30–35 °C for 30 minutes. Then add 1.0 μl more SP6 polymerase (16 units/μl) and continue incubation at 30–35 °C for a further 40 minutes.

2. *Removal of template*:

RNase inhibitor (30 units/μl)	2.0 μl
Total RNA (10 mg/ml)	2.0
DNase I (1 mg/ml)	0.5

Incubate at 37 °C for 10 minutes.

3. *Removal of protein and unincorporated nucleotide*:

Add 1 M DTT	0.8 μl
Sterile dH$_2$O	63.0
3 M NaAc	10.0

a. Extract with 100 μl of phenol/chloroform. Then take aqueous layer (100 μl) and *either* make 2 M with NH$_4$Ac, add 2 volumes ethanol and place on dry ice for 10 minutes *or* pass down a Sephadex G-50 column (Pharmacia; Maniatis et al. 1982). Collect excluded volume, make 2 M with NH$_4$Ac, add 2 volumes ethanol, and place on dry ice for 10 minutes.
b. Microfuge for 5 minutes, dry pellet, and resuspend in 50 μl dH$_2$O made 10 mM DTT.
c. Measure incorporation of radioactive nucleotide in aqueous scintillant.

4. *Alkaline hydrolysis of single-stranded probe*: For efficient in situ hybridization, probes should be only 50–150 bp in length. This is achieved by alkaline hydrolysis of the full-length transcripts with sodium carbonate buffer (pH 10.2), for x minutes at 60 °C. x is dependent on the original length of transcript (Cox et al. 1984). $x = Lo - Lf/KLoLf$ where Lo is the original transcript length in kb; Lf, the final length in kb (i.e., 0.1); K, 0.11; and x, minutes.

5. *Hydrolysis protocol*

   a. Sample in 50 μl of 10 mM DTT.

   b. Add 50 μl of 80 mM $NaHCO_3$, 120 mM Na_2CO_3 (pH 10.2), 10 mM DTT.

   c. Incubate at 60° for x minutes (for 600-nucleotide original transcript, $x = 60$–90 min).

   d. Neutralize with 100 μl of 0.2 M NaAc (pH 6), 1% glacial acetic acid, 10 mM DTT.

   e. Add 20 μl 3 M Na acetate (pH 5.2).

   f. Precipitate with ethanol, microfuge, and dry pellet.

   g. Redissolve in 10 mM DTT (30–60 μl) at a final concentration of 2–3 ng/μl per kilobase.

   h. Check original and final lengths of transcript by electrophoresis in 2% agarose-formaldehyde gels followed by autoradiography (Maniatis et al. 1982).

   i. To the hydrolyzed probe add an equal volume of formamide, and add DTT to a final concentration of 10 mM. This is a 5× probe stock and it should be stored at −20 °C.

G. HYBRIDIZATION

Solutions

1. ³⁵S-labeled probe at 5× concentration in 50% formamide, 10 mᴍ DTT.

2. 10× "Salts": 3 ᴍ NaCl, 0.1 ᴍ Tris-HCl, 0.1 ᴍ NaPO₄ (pH 6.8), 50 mᴍ EDTA, Ficoll 400 (0.2% wt/vol), polyvinylpyrolidone (0.2% wt/vol), BSA Fraction V (0.2% wt/vol). Store at −20°C.

3. Formamide, 2× deionized, 2× recrystallized.

4. 50% Dextran sulfate (Pharmacia). Make up in dH₂O. Heating helps it to dissolve. Store frozen at −20°C.

5. Total RNA 10 mg/ml (see F. Probes).

6. Hybridization buffer. For a final volume of 1 ml, with probe

10× salts	100
Formamide	400
50% Dextran sulfate	200
Total RNA (50 mg/ml)	20
1 ᴍ DTT	8
dH₂O	72
Total volume	800 μl

7. Wash buffer (1× "salts," 50% formamide).

Method

Heat the 5× probe to 80°C for 30 seconds. Microfuge for 20 seconds and mix 1:4 with hybridization buffer, estimating 25 μl per 22×22-mm coverslip. Take slides after air-drying and pipet 25 μl over each group of sections. Gently lower a siliconized, baked coverslip over the sections and incubate overnight at 50°C in a sealed sandwich box containing paper towels saturated with 1× "salts," 50% formamide. If required, the sandwich box can be further sealed within a plastic bag containing 1× salts, 50% formamide.

H. POSTHYBRIDIZATION WASHING

Digestion with RNase A destroys any single-stranded unhybridized probe, while leaving double-stranded hybrids intact. The washing steps are very important to remove background, and variations in time and temperature may be necessary for optimal results. Some probes give higher backgrounds than others.

Solutions

1. Formamide.

2. 10× "salts" (see G. Hybridization).

3. 0.5 M NaCl, in TE (10 mM Tris [pH 7.6], 1 mM EDTA).

4. RNase Sigma Type I-A (Cat. No. R-4875). Make up a stock solution at 10 mg/ml in dH$_2$O, boil for 2 minutes to destroy contaminating nucleases, and store at −20°C. *Note:* Take *GREAT CARE* to keep any pipets, tubes, or containers that have been exposed to RNase separate from other equipment or solutions.

5. 1 M DTT (filter sterilize and store at −20°C).

6. 5 M NH$_4$Ac.

7. 20× SSC (see B. Chemicals and Solutions for Fixation, Sectioning, and Section Pretreatment).

8. Absolute ethanol.

Method

1. Incubate slides in 50% formamide, 1× "salts," 10 mM DTT at 50°C for 1 hour to dislodge the coverslips. Change solution and continue washing with gentle stirring for 1–4 hours.

2. 0.5 M NaCl in TE at 37°C for 15 minutes.

3. 0.5 M NaCl in TE containing 20 µg/ml RNase A at 37°C for 30 minutes.

4. 0.5 M NaCl in TE at 37°C for 30 minutes.

5. Then wash either in 50% formamide/1× "salts" overnight at room temperature at 37°C, or 50°C (depending on how much background is observed in preliminary experiments), or in 3 liters of 2× SSC (30 min at room temperature) followed by 3 liters of 0.1× SSC (30 min at room temperature).

6. Dehydrate through: 30% EtOH in 300 mM NH_4Ac for 2 minutes; 60% EtOH in 300 mM NH_4Ac for 2 minutes; 80% EtOH in 300 mM NH_4Ac for 2 minutes; 95% EtOH in 300 mM NH_4Ac for 2 minutes; 100% EtOH.

7. Air-dry 1–2 hours.

I. AUTORADIOGRAPHY—GENERAL EQUIPMENT AND SOLUTIONS

Darkroom with:

1. 15 W bulb behind Ilford 902S filter.

2. Water bath at 43–45 °C.

3. 5.9 ml of dH$_2$O and 120 μl sterile glycerol premixed.

4. Sterile forceps.

5. Ilford K5 emulsion (fresh). The emulsion should be stored at 4 °C away from penetrating radiation. The longer it is kept, the higher will be the background.

6. 10–20-ml measuring cylinder.

7. Dipping jars (with marks at 6 ml and 12 ml).

8. Sealed Pasteur pipets (sterile).

9. *Clean* unsubbed slides without sections.

10. Diamond pencil.

11. Precooled clean glass or metal plate.

12. Tape, foil, marker, tissues, silica gel.

13. Black exposure boxes with racks. These boxes have silica gel in roof.

14. Large light-proof sandwich box.

15. Timer.

16. Chair.

17. Antiboredom device.

J. AUTORADIOGRAPHY—DIPPING SLIDES

All the steps should be carried out in the darkroom.

1. Warm the 5.9 ml dH$_2$O/120 μl glycerol to 43 °C.

2. Using sterile forceps, put ∽10–15 ml of emulsion shreads into measuring cylinder.

3. Melt at 43 °C.

4. Stir with a sealed Pasteur pipet occasionally at one revolution per second. When molten and smooth, pour down the side of the dipping jar to the 6-ml mark. Pour in 6 ml of H$_2$O/glycerol mix. Stir occasionally at one revolution per second until mixed. Leave 2 minutes. Dip a clean, empty slide into the emulsion, withdraw smoothly, and hold up to safelight. If bubbles are present, continue gentle stirring, leave 2 minutes, and try again. When bubble free, dip an experimental slide, withdraw smoothly, drain 2 seconds, wipe the back of the slide, and put onto precooled level glass or metal plate. Dip all the slides in this way, keeping a check on bubbles. Leave the slides on the precooled plate for 10–15 minutes. Then dry (still horizontal) at room temperature for 60 minutes. (At this stage, you can leave darkroom if there is a seal on light-tight box.) Transfer to exposure box racks, still horizontal, and leave overnight in light-tight exposure boxes containing silica gel at room temperature.

5. Next day put slides to 4 °C.

6. Exposure times are trial and error, but generally range from a few days to a few weeks. Exposure should be monitored by developing individual slides at intervals of a few days.

K. AUTORADIOGRAPHY—DEVELOPING

For full discussion of theory, see Roger (1979). It may take trial and error to find the times in developer and fixer that give the best grain size and background levels. All solutions should be at exactly same temperature (20–22 °C) because variations between solution temperatures cause swelling and contracting of emulsion, leading to cracking and loss of the emulsion layer.

Solutions

1. Kodak D-19 developer.

2. 1% Acetic acid in dH_2O.

3. 30% Sodium thiosulfate.

4. dH_2O.

Method

Warm slides to room temperature. In darkroom take pairs of slides through the following solutions. Be *gentle* and do not agitate.

1. D19: 2–4 minutes.

2. 1% Acetic acid: 1 minute.

3. 30% Sodium thiosulfate: 2–5 minutes (change solution after each pair of slides).

4. dH_2O: 10 minutes.

5. dH_2O: 30 minutes. This can be in the light.

L. STAINING AND MICROSCOPY

Many stains can be used, but Giemsa and Toluidine Blue give nice backgrounds. Suggested: 0.02% Toluidine Blue (filtered through 0.22-μm filter) for 30–60 seconds, depending on emulsion thickness. Be gentle with slides to avoid emulsion cracking.

H_2O	30 seconds
30% EtOH	30 seconds
60% EtOH	30 seconds
80% EtOH	30 seconds
95% EtOH	30 seconds
100% EtOH	30 seconds
Xylene	30 seconds
Xylene	30 seconds

Wipe back of slide, and mount under clean coverslip in mounting fluid (e.g., Permount).

Compare bright-field and dark-ground for best results.

LOCALIZATION OF ANTIGENS IN EMBRYO SECTIONS BY INDIRECT IMMUNOFLUORESCENCE

Again, this is a very basic protocol that may have to be varied depending on the nature of the antigen, the resistance of the epitope to fixation, affinity of the antibody, choice of second antibody, etc.

1. Embryo fixation and embedding. Possible fixation protocols are:
   a. Freshly prepared 4% paraformaldehyde in phosphate-buffered saline (PBS) at 4 °C for 20 minutes to 20 hours, depending on the thickness of the tissue.
   b. A modification of the Sainte-Marie technique (Dziadek and Adamson 1978). Embryos are fixed for 12–24 hours at 4 °C in a mixture of 96% ethanol with glacial acetic acid (99:1 vol/vol).

   After fixation, proceed as for in situ hybridization to the point where cryostat sections are collected onto subbed slides. A good arrangement is five sections per slide, in a staggered pattern.

2. Air-dry sections for 60–120 minutes. Draw around each section individually with a waterproof marker pen. This prevents the antibody solution from spreading.

3. Block unspecific sites with 1% bovine serum albumin (BSA) or 1% nonimmune serum from same species as second antibody in PBS, 30 minutes at room temperature. Place the slide on water-saturated filter paper in a sealed box.

4. Remove solution. Add 50 μl of first antibody in PBS (diluted 1:100–1:1000; microfuge to remove aggregates). Incubate for 30 minutes at room temperature as in step 3.

5. Wash in 500 ml of PBS for 15–60 minutes. A convenient way of doing this is to place the slides in a metal rack in a glass staining jar 4×4×2 in.) full of PBS. Support the rack on a plastic petri dish bottom, inverted over a small magnetic stirring rod. If pieces are cut out of the rim of the petri dish, the PBS will be able to circulate gently when the staining jar is placed on a magnetic stirrer.

6. Dry back of slide and around the sections up to the marks with tissues or filter paper.

7. Add 50 μl fluorescein isothiocyanate (FITC)-labeled second antibody (diluted 1:50–1:100; microfuge to remove aggregates). Incubate for 30 minutes at room temperature, if possible away from light.

8. Wash in 500 ml of PBS for 15 minutes, if possible, away from light.

9. Dry around sections and mount under glass coverslips using semipermanent mounting medium. One medium that gives good results and is based on Rodriguez and Deinhardt (1960) is as follows:

a. 20 g Polyvinyl alcohol (PVA; trade name Gelvatol from Cairn Chemicals Ltd., 60 High Street, Chesham, Bucks, U.K. or Sigma Cat. No. P1763) dissolved over 1–2 days with gentle mixing in 80 ml of 50 mM Tris (pH 8.6) (fluorescein fluoresces optimally at alkaline pH). Add 40 ml of glycerol with gentle mixing for about 4 hours.

b. Centrifuge 10,000 rpm for 15 minutes to remove undissolved PVA and store in small aliquots with minimum air space.

Slides mounted in this way can be stored at 4 °C or at −20 °C, wrapped in aluminum foil. Experts usually add compounds to the mountant that slow down quenching of the FITC by U.V. light, e.g., propyl-gallate (Giloh and Sedat 1982); *para*-phenylenediamine (PPD) (Johnson et al. 1982). However, with these additives the mountant must be stored at 4 °C in the dark and it has a limited shelf life.

Section G

IN VITRO CULTURE
OF EGGS, EMBRYOS, AND
TERATOCARCINOMA CELLS

This section provides details of the culture media and conditions required for isolating eggs and embryos, and for culturing teratocarcinoma cells. The media for collecting eggs and blastocyts for microinjection contain antibiotics and are sterilized by filtration before use. If the embryos are to be returned to the oviduct or uterus after a short incubation period, further precautions against contamination are not required. However, this is not the case if the embryos are to be incubated for more than about 24 hours, when sterile techniques are essential to avoid contamination with bacteria or yeast. In this case, it is convenient to set up the dissection microscope inside a laminar flow hood, or inside a box fitted with a slant-front Perspex panel. Such a box is illustrated in an article by Mintz (1971), which also describes basic embryo culture techniques. Other information about the culture of mouse embryos in vitro is reviewed by Biggers et al. (1971).

GENERAL CONSIDERATIONS

The most important points for successful embryo and cell culture are as follows:

1. Ideally, use disposable sterile plastic dishes, containers, and pipets. Otherwise all glassware used for collecting embryos, pipeting, and storing culture media, and so forth should be absolutely free of detergent residue and rinsed at least six times in glass-distilled water before sterilization.

2. The source of water for making up culture media is extremely important. It should be distilled at least twice in a still in which the metal element is enclosed by glass ($2 \times$ glass-distilled), or be purified by filtration (MilliQ), and stored in clean plastic containers. Prolonged storage is not advised. If necessary, low-level contamination with heavy metals can be overcome by adding 100 μM EDTA to the culture medium (Abramczuk et al. 1977).

3. All chemicals should be of the highest grade possible. It is often necessary to test several batches of a particular component, for example, bovine serum albumin and paraffin oil, and reserve one for culture use. Sources of chemicals used in this section are given in Section H.

4. Do not store the embryo culture media for more than 1 or 2 weeks (see Storage, below). Tissue culture media have a limited shelf life even at 4 °C, particularly if they are made up with glutamine (oxidizes to glutamic acid) and serum (contains glutaminase). If necessary, replenish glutamine before use to a final concentration of 1–2 mM. Storage near a fluorescent light will lead to breakdown of vitamins in culture media.

5. Serum should be aliquoted into small sterile containers and stored at -20 °C. If necessary, it can be heat-inactivated at 56 °C in a water bath for 30 minutes. This is to destroy complement, and the step is not necessary unless the serum contains antibodies against embryo or teratocarcinoma cells. For some experiments batches of serum should be carefully tested for their ability to support cell growth differentiation (see Section E).

6. Most incubations are carried out in a humidified atmosphere of 5% CO_2, 95% air, regulated automatically. A small gassed and humidified container can be placed inside a larger, conventional tissue culture incubator. It has been reported that an atmosphere of 5% CO_2, 5% O_2, and 90% N_2 enhances the survival of early cleavage-stage embryos (Hoppe and Pitts 1973) and this gas phase is preferred by some laboratories (McGrath and Solter 1983b).

7. Cells, and particularly eggs and one- to two-cell cleavage-stage embryos, are intolerant of pH and temperature fluctuations. For some experiments, for example, in vitro fertilization or prolonged manipulation of dissociated blastomeres, it is advisable to use a warm (37 °C) stage for the stereomicroscope. (A stage with preheater can be purchased from Microscope Service and Sales, Egham, Surrey, TW20 9NE, U.K.)

8. Cells will only attach efficiently to tissue culture plastic surfaces. Bacteriological-grade dishes should be clearly labeled to avoid confusion. In addition, tissue culture plastic dishes or inserts are available that are resistant to the chemicals used in histological and electron microscopical procedures (e.g., Falcon 3006 Optical). Microdrop cultures (Section C) should also be made in tissue culture dishes or the drops will not "spread" properly.

9. Protein-containing solutions can be sterilized by positive pressure filtration, for example, through Millipore filters (Type GS, pore size 0.22 μm) or presterilized filter assemblies (Gelman Acrodisc, 0.2-μm pore size). Presterilized Nalgene filter units (Type S 0.20-μm pore size, Nalge Co., Rochester, New York) are convenient for sterilizing up to 200 ml of protein-free solutions. It is wise to check all media batches for sterility before use.

CULTURE MEDIA FOR PREIMPLANTATION-STAGE EMBRYOS

Preimplantation embryos up to the eight-cell stage do not utilize glucose efficiently and they require pyruvate and lactate as energy sources (Iyengar et al. 1983). On the other hand, they do not require amino acids, vitamins, or serum. Bovine serum albumin (BSA) or polyvinylpyrrolidone are usually added to reduce the stickiness of the embryos. The BSA may also serve to adsorb (or supply) trace contaminants (or requirements) in the medium.

Unfortunately, the fertilized eggs of a few inbred mouse strains do not readily continue development beyond the late two-cell stage in vitro. This "two-cell block" can be overcome by injecting cytoplasm from F_1 hybrid embryos that do develop normally (Muggleton-Harris et al. 1982) but the nature of the lesion is not known.

One of the most commonly used culture media is M16, which is very similar to Whitten's medium and is bicarbonate buffered. For collecting embryos, and for experiments in which the embryos are handled for prolonged periods outside the incubator (e.g., microinjection), HEPES buffer is added in place of some of the bicarbonate in order to maintain the correct pH. Medium with HEPES supplementation is known as M2. IMPORTANT: See note on storage on page 255.

Preparation of M16 Culture Medium

M16 is a modified Krebs-Ringer bicarbonate solution and is very similar to Whitten's medium (Whitten 1971). For a full reference on M16 see Whittingham (1971).*

Compound	mм	Molecular weight	g/liter
NaCl	94.66	58.450	5.533[a]
KCl	4.78	74.557	0.356
$CaCl_2 \cdot 2H_2O$	1.71	147.200	0.252
KH_2PO_4	1.19	136.091	0.162
$MgSO_4 \cdot 7H_2O$	1.19	246.500	0.293
$NaHCO_3$	25.00	84.020	2.101
Sodium lactate	23.28	112.100	2.610 or 4.349 g of 60% syrup
Sodium pyruvate	0.33	110.000	0.036
Glucose	5.56	179.860	1.000
Bovine serum albumin (BSA)			4.000
Penicillin G·potassium salt (final conc., 100 units/ml)			0.060
Streptomycin sulfate (final conc., 50 µg/ml)			0.050
Phenol Red			0.010
2× glass-distilled H_2O[b]			up to 1 liter

[a]Increase to 5.68 when $CaCl_2$ omitted for Ca^{++}-free medium.
[b]See note on water under General Considerations.

*Information supplied by Dr. D.G. Whittingham, MRC Experimental Embryology and Teratology Unit, MRC Laboratories, Woodmansterne Road, Carshalton, Surrey SM5 4EF, UK.

PROCEDURE

1. Weigh out penicillin and streptomycin and dissolve in $2\times$ distilled H_2O.

2. Weigh out calcium chloride and dissolve in $2\times$ distilled H_2O.

3. Weigh out remaining compounds (except BSA and lactate) into a 1-liter volumetric flask and add approx 500 ml of $2\times$ distilled H_2O. Allow to dissolve.

4. Add penicillin, streptomycin, and calcium chloride to volumetric flask.

5. Weigh out the lactate syrup into a 10-ml beaker and add to the volumetric flask. Rinse the beaker several times with $2\times$ distilled H_2O, adding the washings to the volumetric flask, and make the volume up to 1 liter.

6. Gas the medium by bubbling 5% CO_2 in air through it for 5 minutes. This step can be omitted if the pH is at 7.4.

7. Sprinkle the BSA on top of the medium and allow to dissolve slowly. Mix gently. Do not shake the medium, as it will froth and denature the protein.

8. Filter through Millipore filter into small sterile containers, gas the air space with 5% CO_2 in air, and cap tightly to maintain a pH of 7.2–7.4. Use positive pressure to filter to reduce foaming. Discard the first few milliliters through the filter.

9. Store at 4°C for up to 1 week. The osmolarity should be 288–292 mosmoles.

Preparation of M2 Culture Medium

This is a modified Krebs-Ringer solution with some of the bicarbonate substituted with HEPES buffer (Quinn et al. 1982). It is used for collecting embryos and for handling them outside the incubator.

Component	mM	Molecular weight	g/liter
NaCl	94.66	58.450	5.533[a]
KCl	4.78	74.557	0.356
CaCl$_2$·2H$_2$O	1.71	147.200	0.252
KH$_2$PO$_4$	1.19	136.091	0.162
MgSO$_4$·7H$_2$O	1.19	246.500	0.293
NaHCO$_3$	4.15	84.020	0.349
HEPES	20.85	238.300	4.969
Sodium lactate	23.28	112.100	2.610 or 4.349 g of 60% syrup
Sodium pyruvate	0.33	110.000	0.036
Glucose	5.56	179.860	1.000
BSA			4.000
Penicillin G·potassium salt			0.060
Streptomycin sulfate			0.050
Phenol Red			0.010
2 × distilled H$_2$O[b]			up to 1 liter

[a]Increase NaCl to 5.68 when CaCl$_2$ omitted for Ca^{++}-free medium.
[b]See note on water under General Considerations.

PROCEDURE

1. Weigh out the HEPES and dissolve in 50–100 ml of 2× distilled H_2O.

2. Adjust the pH to 7.4 with 0.2 N NaOH.

3. Weigh out the penicillin and streptomycin and dissolve in 2× distilled H_2O.

4. Weigh out the calcium chloride and dissolve in 2× distilled H_2O.

5. Weigh out remaining components (except BSA and lactate) into a 1-liter volumetric flask and add ~500 ml of 2× distilled H_2O. Allow to dissolve.

6. Add penicillin, streptomycin, HEPES, and calcium chloride to the volumetric flask.

7. Weigh out the lactate syrup into a 10-ml beaker and add to the volumetric flask. Rinse the beaker several times with 2× distilled H_2O, adding the washings to the volumetric flask and making the volume up to 1 liter.

8. Finally, sprinkle BSA on top of the medium and allow to dissolve slowly. Mix gently. Do not shake the medium as it will froth and denature the protein.

9. If necessary, readjust the pH of the medium to 7.2–7.4 with 0.2 N NaOH.

10. Filter through Millipore filter with positive pressure, discarding first few milliliters and aliquot into sterile containers.

11. Store at 4°C for up to 1 week. The osmolarity should be 285–287 mosmol.

Preparation of M2 and M16 from Concentrated Stocks

It is often more convenient to prepare concentrated stocks of the components of M2 and M16, but take note of the information on storage below.

Stock A	Component	g/100 ml
(10× concentration)	NaCl	5.534
	KCl	0.356
	KH_2PO_4	0.162
	$MgSO_4 \cdot 7H_2O$	0.293
	sodium lactate	2.610
		or 4.349 g of 60% syrup
	glucose	1.000
	penicillin	0.060
	streptomycin	0.050

Stock B	Component	g/100 ml
(10× concentration)	$NaHCO_3$	2.101
	phenol Red	0.010

Stock C	Component	g/10 ml
(100× concentration)	sodium pyruvate	0.036

Stock D	Component	g/10 ml
(100× concentration)	$CaCl_2 \cdot 2H_2O$	0.252

Stock E	Component	g/100 ml
(10× concentration)	HEPES	5.958
	Phenol Red	0.010

Take 5ml stock soln add water to 50 ml.

PROCEDURES

Stock A

1. Weigh out salts (except sodium lactate) into a volumetric flask.

2. Weigh out sodium lactate into a beaker.

3. Add the sodium lactate to the volumetric flask.

4. Rinse the beaker several times with $2\times$ distilled H_2O, adding the washings to the flask.

5. Make up to volume using $2\times$ distilled H_2O.

Stocks B, C, and D

1. Weigh out salts into volumetric flask, and make up to volume using $2\times$ distilled H_2O.

Stock E

1. Weight out HEPES and Phenol Red into a beaker.

2. Add approx 50 ml of $2\times$ distilled H_2O and allow to dissolve.

3. Adjust the pH to 7.4 with 0.2 N NaOH.

4. Pour into a 100-ml volumetric flask.

5. Rinse the beaker with $2\times$ distilled H_2O, add washings to the volumetric flask, and make up to 100 ml.

6. Filter stocks through Millipore filter into sterile plastic tubes.

Storage

If stored in a refrigerator at 4°C, stocks A, D, and E can be kept up to 3 months, but it is important that stocks B and C are changed every other week. Stocks stored frozen at −20°C can be kept for longer periods, but once the $1\times$ solution is made up it should not be stored for more than 1 week.

M2 from Concentrated Stocks

Stock	10 ml	50 ml	100 ml	150 ml	200 ml
A($\times$10)	1.00	5.0	10.0	15.0	20.0
B($\times$10)	0.16	0.8	1.6	2.4	3.2
C($\times$100)	0.10	0.5	1.0	1.5	2.0
D($\times$100)	0.10	0.5	1.0	1.5	2.0
E($\times$10)	0.84	4.2	8.4	12.6	16.8
H_2O[a]	7.80	39.0	78.0	117.0	156.0
BSA	40 mg	200 mg	400 mg	600 mg	800 mg

[a]See note on water under General Considerations.

1. Measure out 2$\times$ distilled H_2O accurately into a conical flask.

2. Measure out stock solutions using plastic pipet or tips, and leave pipet in the conical flask.

3. Rinse the pipets in the flask by sucking up water/medium mixture two to three times.

4. Measure the osmolarity of the medium. (This step is optional.)

5. Add the BSA to the medium to a final concentration of 4 mg/ml, allow to dissolve slowly, and mix gently.

6. If necessary readjust the pH of the medium with 0.2 N NaOH to pH 7.2–7.4, using color standards.

7. Millipore filter into sterile plastic tubes.

8. Store at 4°C for up to 1 week.

M16 from Concentrated Stocks

Stock	10 ml	50 ml	100 ml	150 ml	200 ml
A (×10)	1.0	5.0	10	15.0	20
B (×10)	1.0	5.0	10	15.0	20
C (×100)	0.1	0.5	1	1.5	2
D (×100)	0.1	0.5	1	1.5	2
H_2O[a]	7.8	39.0	78	117.0	156
BSA	40 mg	200 mg	400 mg	600 mg	800 mg

[a]See note on water under General Considerations.

1. Measure out 2× distilled H_2O accurately into a conical flask.

2. Measure out stock solutions using plastic pipets, and leave pipet in the conical flask.

3. Rinse the pipets in the flask by sucking up water/medium mixture two to three times.

4. Measure osmolarity of the medium. (This step is optional.)

5. Gas the medium with 5% CO_2 in air for approx 15 minutes to adjust pH to 7.4.

6. Add the BSA to give 4 mg/ml, allow to dissolve slowly, and mix gently.

7. Millipore filter into small glass bottles or plastic tubes, gassing the space with 5% CO_2 in air for ~30 seconds and cap tightly to maintain a pH of 7.2–7.4.

8. Store at 4°C for up to 1 week.

Other Media Used for Culturing Preimplantation Embryos

Mintz (1964) reports using 50% fetal bovine serum and 50% Earle's salt solution containing 0.002% Phenol Red, and lactic acid at 1.0 mg/ml, for aggregation chimera formation and culture. The pH is adjusted to 7.0 with 7.5% $NaHCO_3$ and the embryos cultured in an atmosphere of 5% CO_2, 95% air.

Reference to the medium used by Hoppe can be found in Hoppe and Pitts (1973). It is very similar to Whitten's medium and to M16.

MEDIUM FOR IN VITRO FERTILIZATION OF MOUSE OOCYTES

Use of this medium supplemented with 30 mg/ml BSA for in vitro fertilization is described in Section C (In Vitro Fertilization) (Whittingham 1971, modified by L. Fraser; Fraser and Drury 1975).

Component	g/100 ml[a]
NaCl	0.5803 g
NaHCO$_3$	0.2106
Glucose	0.1000
KCl	0.0201
Na$_2$HPO$_4$12H$_2$O	0.0056
Sodium pyruvate	0.0055
Penicillin	0.0063
Streptomycin	0.0050
CaCl$_2$2H$_2$O	0.0264
MgCl$_2$6H$_2$O	0.0102
Sodium lactate	0.35 ml of 60% syrup
Phenol Red	0.1000

[a]See note on water under General Considerations.

Make up to 100 ml with distilled water, Millipore filter, then aliquot in 10 × 10 ml and store frozen.

CULTURE MEDIA FOR BLASTOCYSTS, POSTIMPLANTATION EMBRYOS, AND TISSUES

By the late morula stage, embryos no longer require pyruvate and lactate but can efficiently utilize glucose as an energy source. However, they now have more complex requirements in terms of amino acids, vitamins, and serum. A few defined media are available in which the serum is replaced with growth factors, but a complete analysis of the growth factors produced and required by early embryo cells is not yet available.

The following formulated media have been used successfully for culturing postimplantation embryos, when supplemented with serum: (1) CMRL 1066 (Gonda and Hsu 1980); (2) Eagle's minimal essential medium (MEM) (Pienkowski et al. 1974; Solter and Knowles 1975); and (3) Dulbecco's modified Eagle's medium (DMEM) (Hogan and Tilly 1978a,b; Tam and Snow 1980; Evans and Kaufman 1981, 1983; Martin 1981).

For radioactive labeling with [^{35}S]methionine, use either methionine-free culture medium (for incubation periods of up to 1 hr) or medium with 1 μg/ml cold methionine (for up to 16 hr), in either case with dialyzed serum. Cold methionine is required for longer incubations so that the cells do not become starved of this essential amino acid.

Commercial fetal bovine serum at 10–20% is most commonly used for short-term cultures of isolated embryonic pieces, but, as described below, optional growth, differentiation, and morphogenesis of more intact embryos require higher concentrations of carefully prepared and handled serum.

Culture of Blastocysts

Procedures for the culture of blastocysts attached to tissue culture dishes have been described in detail by Hsu and his collaborators (Hsu 1978, 1980; Gonda and Hsu 1980). They have found the best development is obtained when up to 40% (vol/vol) human placental cord serum is added to the culture medium, which is changed frequently. Under these conditions 60% of blastocysts develop to the early somite–neural fold stage (Wu et al. 1981).

Culture of Primitive Streak/Early Somite-stage Embryos

Tam and Snow (1980) have made a detailed, comparative study of the culture conditions giving good development of early primitive streak-stage mouse embryos incubated in tissue culture dishes. Under optimal conditions they found that 50% of their primitive streak egg cylinders would progress to early somite and beating heart stages. The best medium consisted of DMEM, diluted 50% (vol/vol) with mouse serum prepared under careful conditions. Full details are given in Tam and Snow (1980). Briefly, it is important that the medium contains a high concentration of glucose (as in the formulation of DMEM used) and is freshly supplemented with glutamine (final concentration 2 mm) and sodium pyruvate (final concentration

0.1 mm). The mouse serum is prepared from blood centrifuged immediately after collection and not allowed to clot. It should not be heat-inactivated nor filter-sterilized before storage.

Conditions for culturing early somite-stage rat embryos in roller bottles have been described by New (1978). A high proportion of these embryos undergo neural tube closure. Similar conditions can be used to study this process in mouse embryos.

Preparation of Human Placental Cord Serum

To collect blood from the human placenta and umbilical cord, the first step is to establish friendly relations with junior doctors and particularly the nursing staff of the maternity department of the local hospital. Explain that you require blood from placentas as soon after delivery as possible and ask the senior nurse on duty to call you when a birth is imminent. In practice she or he is usually too busy to phone, so it is advisable to call during the day to check. The placenta has to be inspected and weighed by the nurse, so expect some delay. Hold up the placenta and clamp the cord in two places (Fig. 76). Squeeze blood from the placenta down the cord and then cut between the clamps. Hold the cut end over a clean, 10-ml *glass* container. Release the clamp and collect blood while continuing to squeeze placenta. Allow blood to clot at 4°C. In glass tubes the clot will not adhere to the sides. Store at −20°C or −70°C. Expect about 10 ml of serum from one placenta.

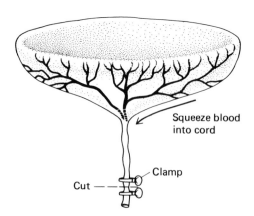

Figure 76 Diagram illustrating method for obtaining cord blood from a human placenta.

Serum-free Defined Media for Embryo Culture

A serum-free defined medium has been used for blastocyst culture, but the extent of development of these embryos is very limited at present (Rizzino and Sherman 1979). Martin (1981) describes the use of Dulbecco's modified Eagle's medium (DMEM) supplemented with 10 μg/ml insulin and 5 μg/ml transferrin for producing conditioned medium of PSA1 pluripotent teratocarcinoma cells. After concentration by lyophilization, this medium is supplemented with fresh DMEM and 10^{-4} M β-mercaptoethanol for isolating pluripotential cell lines from blastocysts.

CULTURE OF TERATOCARCINOMA CELLS

F9 Teratocarcinoma Cells

F9 teratocarcinoma cells were established in culture from a subline of the transplantable tumor OTT6050, itself derived from a 6-day male embryo transplanted into the testis (see Bernstine et al. 1973). At first the cells gave rise to well-differentiated tumors in vivo, but in most laboratories the spontaneous rate of differentiation both in vivo and in vitro is now very low. However, when treated with low concentrations of retinoic acid, F9 stem cells will differentiate into either parietal endoderm or visceral endoderm-like cells, depending on the culture conditions (Strickland and Mahdavi 1978; Strickland et al. 1980; Hogan et al. 1981; for review, see Hogan et al. 1983). The undifferentiated embryonal carcinoma (EC) cells grow well in Dulbecco's modified Eagle's medium (DMEM) supplemented with 10% fetal bovine serum. The morphology of the undifferentiated cells is shown in Figure 77, A and B. They adhere poorly to untreated tissue culture plastic dishes, but will attach more firmly to dishes treated with gelatin. EC cells do not become quiescent at high density or low serum concentrations but lyse. They must therefore be subcultured frequently. Stocks should be kept frozen in liquid nitrogen. Few accurate and recent karyotypes of F9 clonal lines are available. Dr. Liz Robertson of the Department of Genetics, University of Cambridge has found that all her clones have 39 chromosomes and are XO with monosomy of 14, trisomy of 8, and a Robertsonian translocation (2.8) (pers. comm.).

Preparation of Gelatin-coated Tissue Culture Dishes

Pipet a sterile 1% solution of gelatin (see Section H, Chemicals) in 2× glass-distilled water into each dish to cover the surface. After a few minutes remove most of the solution (it can be used to treat other dishes), leaving a thin film on the surface. The dishes can be used immediately or left to dry at room temperature, after which they may be stored for several months.

Subculture with Trypsin/EDTA

Dilute a stock solution of trypsin (Section H, Solutions) 1:4 into 0.02% EDTA/saline (Section H, Solutions) to give a final trypsin concentration of 0.05%. Remove the culture medium from the cells, rinse twice with 0.05% trypsin/EDTA, and then cover the cells with a minimal volume of the trypsin/EDTA solution (approximately 1 ml/9-cm dish). Incubate at 37°C for a few minutes until the cells have rounded up and detached. Quickly add fresh medium with serum to inhibit the trypsin and seed cells into fresh plates at a density of approximately 3×10^5 cells/90-mm dish or 2×10^4 cells/35-mm dish. Incubate at 37°C in a humidifed atmosphere of 5% CO_2, 95% air.

Note that the trypsin used for cell culture is only crude. For biochemical studies, for example on cell-surface components, purified trypsin (Sigma bovine pancreas Type III, Cat. No. T8253) and soybean trypsin inhibitor should be used.

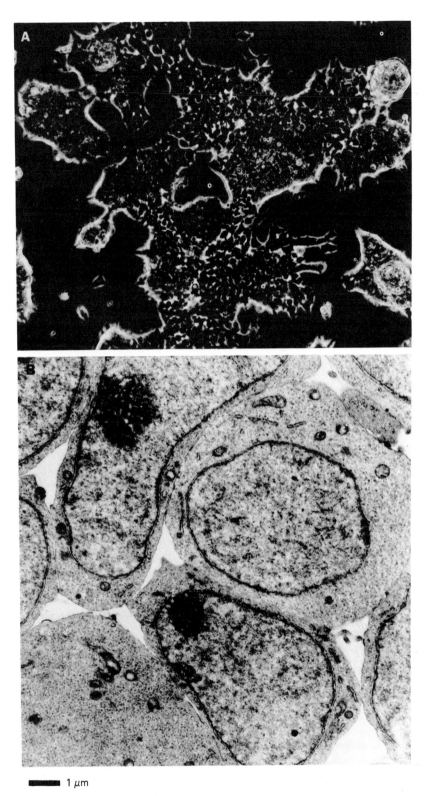

1 μm

Figure 77 Undifferentiated F9 teratocarcinoma cells in culture. (*A*) Phase-control photomicrograph of cells growing as a monolayer on gelatin-coated tissue culture dish. (*B*) Transmission electron micrograph showing absence of specialized intercellular junctional complexes and poorly developed endoplasmic reticulum.

Differentiation into Parietal Endoderm-like Cells

F9 EC cells are seeded at low density (approximately 3×10^5/90-mm-diameter dish or 1×10^6/15-mm-diameter dish) in DMEM with 10% fetal bovine serum containing 5×10^{-8} M retinoic acid, 10^{-4} M dibutyryl cAMP, and 10^{-4} M isobutyl methyl xanthine. The medium should be replaced every 48 hours. The cells will also differentiate into parietal endoderm-like cells if they are treated with retinoic acid first and then cAMP, but cAMP alone has little or no effect (Strickland et al. 1980) (see Fig. 78A–C).

Source	Stock solution	Comments
Trans-retinoic acid (Sigma)	10^{-4} M in ethanol; assumed sterile	store solid in the dark at $-20\,°C$; store stock solution in the dark at $4°$ for not more than 1 week; protect solutions and cultures from fluorescent light
Dibutyryl cAMP (Sigma)	10^{-1} M in DMEM or H_2O; filter sterilize	store stock solution at $4\,°C$
Isobutylmethyl-xanthine (phosphodiesterase inhibitor) (Sigma)	dissolve in DMSO to 10^{-1} M; dilute to 10^{-2} M in DMEM or H_2O; filter sterilize	store stock solution at $4\,°C$

Figure 78 Differentiation of F9 cells into parietal endoderm. (A) Control monolayer culture of F9 cells; (B) 4 days after adding 1×10^{-7} M retinoic acid the cells have become flattened; (C) 4 days after adding 1×10^{-7} M retinoic acid and 10^{-3} M dibutyryl cAMP, the cells are more rounded and are secreting large amounts of extracellular matrix material including type-IV collagen and laminin. A similar response is seen with 5×10^{-8} M retinoic acid, 10^{-4} M dibutyryl cAMP, and 10^{-4} M isobutylmethyl xanthine.

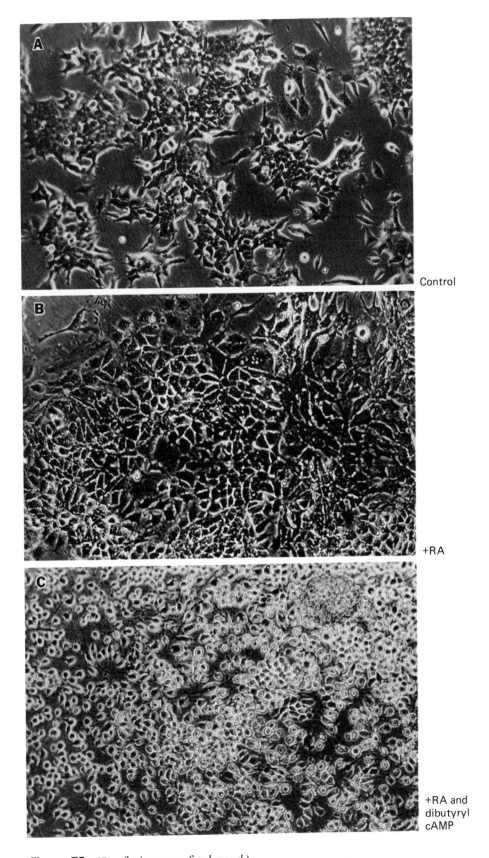

Control

+RA

+RA and
dibutyryl
cAMP

Figure 78 (*See facing page for legend.*)

Differentiation into Visceral Endoderm-like Cells

F9 cells are trypsinized briefly and gently pipeted so that they disperse into small clumps, each of about 20 cells. These clumps are seeded into bacteriological plastic petri dishes (to which the cells do not adhere) in DMEM containing 10% fetal bovine serum and 5×10^{-8} M retinoic acid. Under these conditions the aggregates will differentiate into embryoid bodies with an outer epithelial layer of visceral endoderm-like cells synthesizing α-fetoprotein and transferrin (Hogan et al. 1981; Adamson and Hogan 1984; Scott et al. 1984). As the aggregates enlarge, the cultures should be either split or "thinned out" to prevent overcrowding. During this subcloning, the aggregates should be handled gently using a wide-bore (agar) pipet to avoid damaging the outer epithelial layer. It is not necessary to add fresh retinoic acid when subculturing the aggregates (see Fig. 79A,B).

Other Teratocarcinoma Cell Lines

A wide variety of pluripotent and nullipotent teratocarcinoma cell lines are available (see Silver et al. 1983 for details). One useful EC cell line is P19, described by McBurney et al. (1982). This line differentiates into muscle, neuronal cells, and extraembryonic endoderm, depending on the inducer (DMSO or retinoic acid) and culture conditions.

Serum-free Media for Teratocarcinoma Cells

For a review of recent studies on the growth factor requirements of differentiated and undifferentiated teratocarcinoma cells, see Heath (1983). Serum-free defined media are discussed in this review and details can also be found in Darmon et al. (1981) and Rizzino and Crowley (1980).

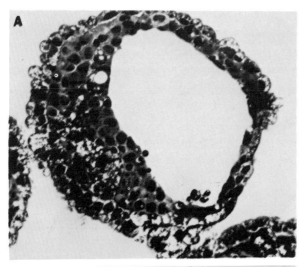

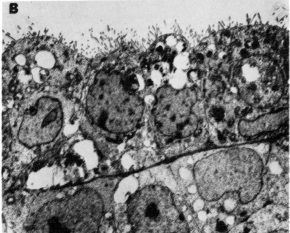

Figure 79 Differentiation of F9 cells into visceral endoderm. (A) Section of an "embryoid body" formed 7 days after treating an aggreate of F9 cells with 5×10^{-8} M retinoic acid. The outer epithelial layer of endoderm cells synthesizes α-fetoprotein, and surrounds an inner layer of undifferentiated cells. (B) Transmission electron micrograph of the visceral endoderm cells, showing apical microvilli and desmosomal junctions, numerous vacuoles, and thin basal lamina.

Section H

CHEMICALS, SUPPLIES, AND SOLUTIONS

This section lists some of the sources of chemicals, instruments, and equipment descibed in the Manual. Alternative sources are, of course, possible, and in any case care should be taken to test all new batches of components of embryo culture media, hormones, etc., before large-scale use.

CHEMICALS AND SUPPLIES

Agarose

Sigma Type 1, low EEO (Cat No. A-6013).

Avertin

A stock of 100% Avertin is prepared by mixing 10 g of tribromoethyl alcohol with 10 ml of tertiary amyl alcohol. To use, dilute to 2.5% in water or isotonic saline. Both the 100% and 2.5% stocks are stored wrapped in foil at 4 °C. The proper dose of Avertin may vary with different preparations and should be redetermined each time a new 100% stock is made; it will be around 0.015–0.017 ml/g body weight.

Bovine Serum Albumin

May be either powdered (Fraction V) Sigma Cat. No. A-9647 or crystalline (e.g., Pentex BSA Code No. 81-001 from Miles Laboratories Ltd). Some batches of Fraction V may contain spermine oxidase, which inhibits development beyond the eight-cell stage (D.G. Whittingham, pers. comm.). Some labs use Sigma Cat. No. A-4378. All batches should be tested for toxicity before use.

Desmarres Chalazion Forceps 26-mm Diameter

Joseph E. Frankle C., Laboratory Supplies and Equipment. 4309-11 Rising Sun Ave., Philadelphia, Pennsylvania 19140 (Cat. No. C 62-2810).

Dibutyryl cAMP

Sigma.

Gelatin

Sigma swine skin Type I (Cat. No. G-2500).

Glucose

Analar BDH.

HEPES buffer

Sigma (Cat. No. H-3375).

Hoffman Optics

Hoffman Modulation Optics Inc., 100 Forest Drive at East Hills, Greenvale, New York 11548. Supplier in the U.K. is Microinstruments (Oxford) Ltd., 7 Little Clarendon St., Oxford OX1 2HP.

Hormones for Superovulation

Pregnant mare's serum gonadotropin (PMS) is sold under the commercial names Gesty L (Organon Inc., West Orange, New Jersey 07052), Folligon (Intervet Laboratories Ltd., Science Park, Milton Road, Cambridge CB4 4DJ, UK), or serum gonadotropin (Sigma, Cat. No. G 4877).

Human chorionic gonadotropin (hCG) is known by the commercial names Chorulon (Intervet Labs) or Pregnyl (Organon Inc.).

Preparation and storage are described in Section C.

Hyaluronidase

Hyaluronidase used is Type IV-S from bovine testes, Sigma Cat. No. H3884. Prepare as a stock solution at 10 mg/ml in M2 medium or H_2O. Filter-sterilize, aliquot, and store at $-20°C$. Dilute to ∼300 μg/ml in M2 with BSA for removing cumulus cells (see Section C).

Isobutyl Methyl Xanthine

Sigma.

Micromanipulators

Narashige equipment is marketed in the United States through Medical Systems Corp., 239 Great Neck Road, Great Neck, New York 11021. A range of micromanipulators from various sources is supplied by Microinstruments (Oxford) Ltd. (for address, see Hoffman Optics, above).

Needle Holders

Tungsten loop holders as used in bacteriological work are available from Gallenkamp (DKD 430N) and Scientific Supplies (W1740).

Oil

A light liquid paraffin oil; e.g., Fisher 0-119; BDH Liquid Paraffin (light), weight per ml at 20°C 0.83–0.86 (Prod. 29436. BDH Chemicals Ltd., Poole, England); or paraffin oil sold by chemists (pharmacists) for human medicinal use. Some investigators sterilize the oil by autoclaving. However, other investigators believe that this greatly increases toxicity and is not necessary since cultures containing antibiotics rarely become infected. In either case, the potential toxicity of batches of oil should be checked on spare embryos.

Ophthalmic Suture Needles

No. 6 eye, half-curved can be obtained from the Holborn Surgical and Veterinary Instruments Ltd. Dolphin Works, Margate Road, Broadstairs, Kent, U.K. (Cat No. E705).

Pancreatin

Porcine from BDH or Sigma Cat. No. P-1500.

Penicillin G

Benzylpenicillin, K salt Sigma, approximately 1600 units/mg, or Glaxo.

Polyvinylpyrrolidone

Polyvinylpyrrolidone (PVP) is available from Sigma (m.w. 40,000, PVP-40) and BDH (m.w. 44,000, BDH Cat. No. 29579).

Pronase

Protease from *Streptomyces griseus* available from Calbiochem, Boehringer (165 921), or Sigma (P-5147). For removing zonae prepare as a 0.5% solution in M2. If necessary 0.5% PVP can be added to reduce the stickiness of the embryos. Since Pronase is a crude enzyme preparation, it probably should be incubated for 30 minutes at room temperature to destroy contaminating nucleases, etc. Centrifuge to remove insoluble material, filter sterilize, and store in aliquots at $-20°C$.

Retinoic Acid—All trans

Sigma.

Sodium Lactate

DL lactic acid, Na salt, 60% syrup (Sigma, Cat. No. L 1375).

Sodium Pyruvate

Pyruvic acid (Na salt) Type II (Sigma, Cat. No. P2256, or Calbiochem).

Streptomycin Sulfate

Sigma S-6501 or Glaxo.

Surgical Instruments

Roboz Surgical Instrument Co., Inc., 1000 Connecticut Ave., N.W., Washington, D.C. 20036. In United Kingdom, Holborn Surgical and Veterinary Instruments Ltd, Dolphin Works, Margate Road, Broadstairs, Kent.

Trypsin

Bovine pancreas Type III, Sigma Cat. No. T8253, 2× crystallized for dissecting tissue layers, or Difco 1:250 for tissue culture subculturing.

Wire: Platinum

Platinum wire for replacing filaments of the microforge can be obtained from Goodfellow Metals (see below).

Wire: Tungsten

99.95% pure, 0.5-mm diameter. Available from Goodfellow Metals, Science Park, Milton Road, Cambridge CB4 4DJ, U.K. (Cat. No. W005160/4); Ernest F. Fullam, Inc, P.O. Box 444, Schenectady, New York 12301 (Cat. No. 16210). For sharpening tungsten needles the apparatus described in Figure 80 is very convenient. A very small piece of plasticine can be attached to the tip of the blunt needle. Stop the electrolysis when it drops off.

Wound Clips

Clay Adams autoclips (9 mm), autoclip applier, and autoclip remover from Arthur H. Thomas Co., Vine Street at 3rd, P.O. Box 779, Philadelphia, Pennsylvania 19105. In United Kingdom, available from Arnold R. Horwell Ltd, 73 Maygrove Road, West Hampstead, London NWG 2BP.

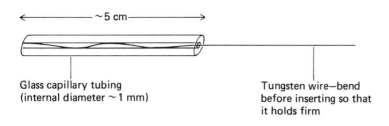

Glass capillary tubing
(internal diameter ~1 mm)

Tungsten wire—bend
before inserting so that
it holds firm

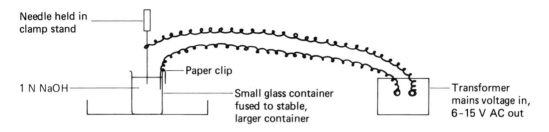

Needle held in
clamp stand

Paper clip

1 N NaOH

Small glass container
fused to stable,
larger container

Transformer
mains voltage in,
6–15 V AC out

Figure 80 Apparatus for sharpening tungsten needles by electrolysis.

SOLUTIONS

All solutions should be made up in 2× glass-distilled water and stored in disposable plastic or clean glass containers absolutely free of detergent residue.

Acidic Tyrode Solution for Removing Zonae

	g/100 ml
NaCl	0.800
KCl	0.020
$CaCl_2 \cdot 2H_2O$	0.024
$MgCl_2 \cdot 6H_2O$	0.010
Glucose	0.100
Polyvinylpyrrolidone (PVP)	0.400

Adjust to pH 2.5 with Analar HCl. The PVP is added to increase the viscosity and reduce embryo stickiness. Sterilize by filtration. Store in aliquots at −20 °C (stocks) or room temperature.

Saline/EDTA Solution for Isolation of Germ Cells and Tissue Culture.

	g/100 ml in 2× distilled H_2O
EDTA (disodium salt, Sigma)	0.02
NaCl	0.80
KCl	0.02
Na_2HPO_4 (anhydrous)	0.115
KH_2PO_4	0.02
Glucose	0.02
Phenol Red	0.001

The final EDTA concentration is 0.02%. Check that pH is 7.2. Sterilize by filtration or by autoclaving (121 °C, 15 lbs. p.s.i.) for 15 minutes. Store at room temperature. For tissue culture alone, the glucose can be omitted.

0.25% Trypsin in Tris-saline for Tissue Culture

	g/liter
NaCl	8.00
KCl	0.40
Na_2HPO_4	0.10
Glucose	1.00
Trizma base (Sigma)	3.00
Phenol Red	0.010
Penicillin G	0.060 (final conc., 100 units/ml)
Streptomycin	0.100
Trypsin (Difco 1:250)	2.5 g (dissolve in small volume of water before adding)

Trypsin stock solution for tissue culture 0.25% trypsin Difco 1:250 in Tris-saline (or any other well-buffered isotonic salt solution). Addition of antibiotics is optional. Adjust pH to $\sim$ 7.6. Filter-sterilize and aliquot into sterile containers. Store at $-20\,°C$. This stock is diluted 1:4 in saline/EDTA before use.

Pancreatin/Trypsin Solution for Separating Tissue Layers

	g/20 ml	Final concentration
Pancreatin	0.50	2.5%
Trypsin	0.10	0.5%
Polyvinylpyrolidone	0.10	0.5%

Modified from Levak-Svajger et al. (1969). Make up in DMEM (or similar buffered salt solution with glucose and amino acids) without Ca^{++} and with 20 mM HEPES (pH 7.4). The suspension will be difficult to filter-sterilize without low-speed centrifugation or prefiltering. Store sterile in small aliquots at $-20\,°C$.

Phosphate-buffered Saline pH 7.2

	g/100 ml
NaCl	0.800
KCl	0.020
Na_2HPO_4	0.115 or $Na_2HPO_4·12HO$ 0.289
KH_2PO_4	0.020

This solution can also be obtained in tablet form and is the same as PBSA of Dulbecco. Millipore-filter and store at $4\,°C$. It can also be autoclaved at 15 lb. pressure for 15 minutes. If required, $CaCl_2H_2O$ and $MgCl_2·H_2O$ can be added at 0.014 g/100 ml and 0.010 g/100 ml, respectively. This is "PBS complete" or solution PBSABC of Dulbecco.

Appendix 1:

Platforms and Baseplate for Nikon- or Zeiss-based Microinjection Apparatus

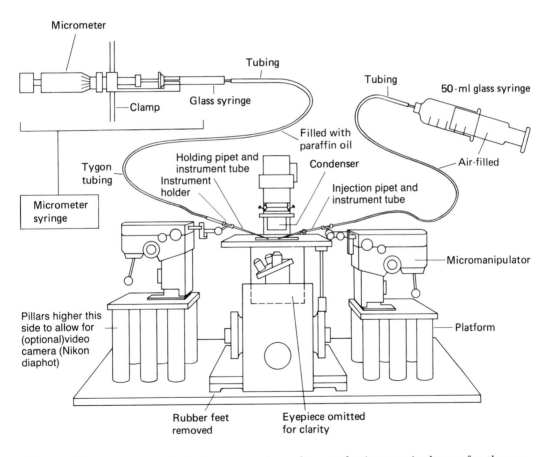

Figure A1 Arrangement of microscope, baseplate, and micromanipulators for the pronuclear injection methods described in this manual. Details of the baseplate and platforms are given in this Appendix, and sources of the apparatus are listed in Section D.

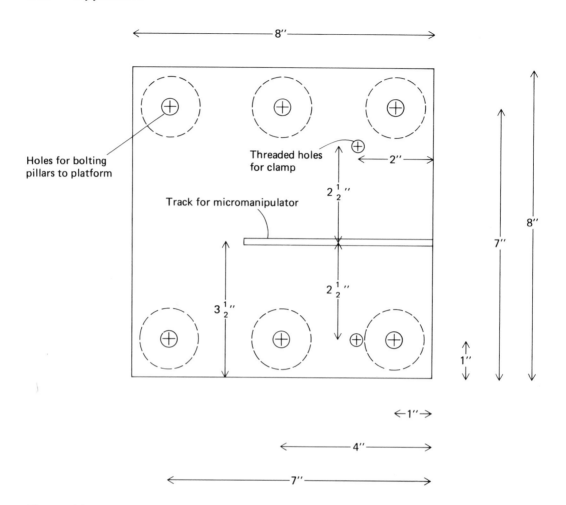

Figure A2 The platform is a ¾-in. plate with rounded corners and beveled edges. The pillars are 1½-in.-diameter aluminum rods that are 5½ in. long. The left pillars for the TV camera should be 7 in. tall.

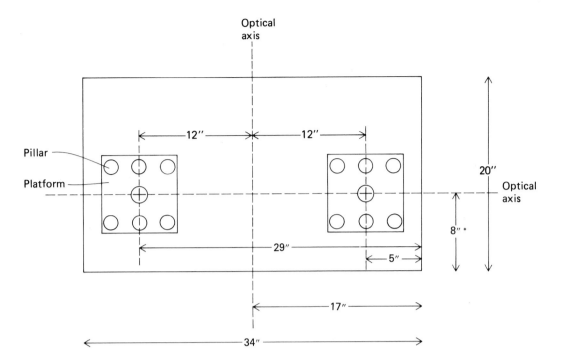

Figure A3 Diagram of baseplate for Nikon Diaphot showing the location of the two platforms. Machinist's notes on baseplate construction: 1. The 12 pillars are 1½ in. in diameter × 5½ in. long; each is drilled and tapped at each end for ¼-20 × ¾ in. deep. 2. Twelve socket head cap screws ¼-20 × 1.0 in. are required to mount the 12 pillars to the baseplate. 3. To adapt for a TV camera, use six pillar extensions 1½ in. diameter × 1½ in. long, which are drilled and tapped at each end for ¼-20 × ¾ in. deep. (An alternative is six 1½ in.-diameter × 7.0-in.-long pillars in lieu of spacers.) 4. Eighteen threaded studs ¼-20 × 1.0 in. long are required to mount pillars to platform plates and extensions to pillars. 5. The six holes in platforms to mount pillars are ¼-20 × ½ in. to ⅝ in. deep. 6. Two holes for clamps in platform are ¼-20 × ½ in. deep on the side opposite holes for mounting pillars. For Zeiss IM-35, the pillars should be 5 in. long (not 5½ in.) and the platforms should be placed so that the optical axis is 9 in. (not 8 in.) from the front of the baseplate. (*) For Zeiss IM-35, this should be 9 in.

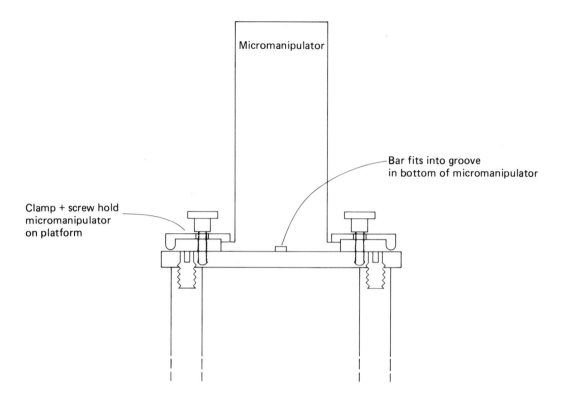

Figure A4 Schematic front view of micromanipulator setup.

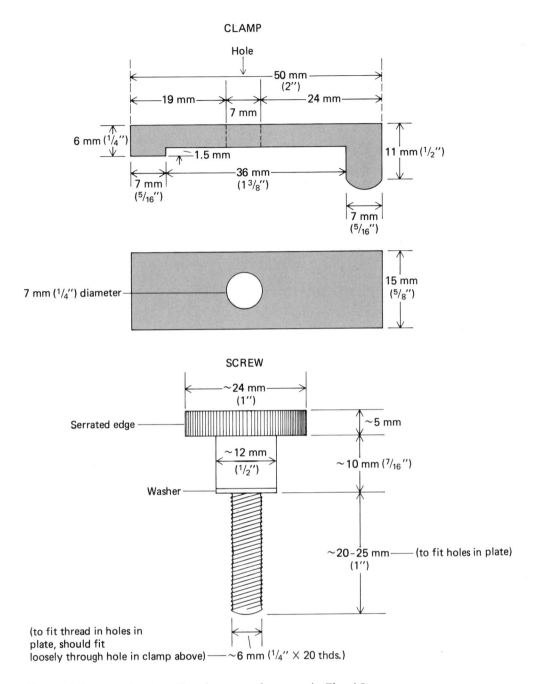

Figure A5 Specifications for clamps and screws in Fig. A3.

Information supplied by Dr. Douglas Hanahan, Cold Spring Harbor Laboratory, Cold Spring Harbor, New York 11724.

Appendix 2:
Linkage Map of the Mouse

Muriel T. Davisson and Thomas H. Roderick
The Jackson Laboratory
Bar Harbor, Maine 04609

The linkage map of the mouse is a graphic summary of the efforts of many investigators at many institutions and represents data accumulated since the first linkage group was described in the mouse in 1915 (Haldane et al. 1915). Because the map includes different kinds of data, we have developed several conventions (described below) to depict as accurately as possible the current status of gene mapping in the mouse.

The solid vertical bars represent the chromosomes of the mouse and are drawn to their proportional lengths based on an estimated total haploid length of 1600 cM. Chromosome numbers are shown above the centromeres, which are represented by knobs. When observed genetic lengths for chromosomes exceed expected lengths, the vertical bars are extended by hatched lines. Gene symbols are at the right of the chromosome bars. Nucleolus organizers are symbolized by NO, and tightly linked complexes of related genes are represented by one symbol. Gene symbols and names are listed alphabetically in the table following the map.

Numbers to the left of the chromosomes are recombination percentages. The values represent data from different kinds of linkage crosses as well as from recombinant inbred strains. The map is compiled from experiments where either females or males were used as the heterozygous parent. A table of recombination percentages with references for much of these data appears in Davisson and Roderick (1981). Distances between centromeres and proximal markers, when determined by using Robertsonian chromosomes, are enclosed in parentheses because these distances may be underestimated. When the distance between the first proximal marker and a centromere is not known, the linkage group is centered on the chromosome. Genes that have been shown to be syntenic with specific chromosomes by parasexual methods are listed at the bottom of their respective chromosomes.

Further conventions are used to show the relative certainty of position of a locus and the relationships of loci. The longer the line through the chromosome, the more certain is the position. Positions of some genes are well known from three-point crosses and extensive data. These "anchor loci" are indicated by lines extending through the vertical chromosome bars and beyond all others to the left. When the position of a locus is known with respect to an adjacent locus, but has not been tested with loci further away, the line is drawn to, but not through, the chromosome. Symbols for genes found to be in tandem by molecular mapping extend out from the same point on the chromosome, e.g., *Alb-1* and *Afp* on chromosome 5. Genes whose order is known with respect to at least two other loci are italicized. Nonitalicized loci have been mapped with respect to only one other locus and, hence, could be located on either side of that locus. The symbol of the

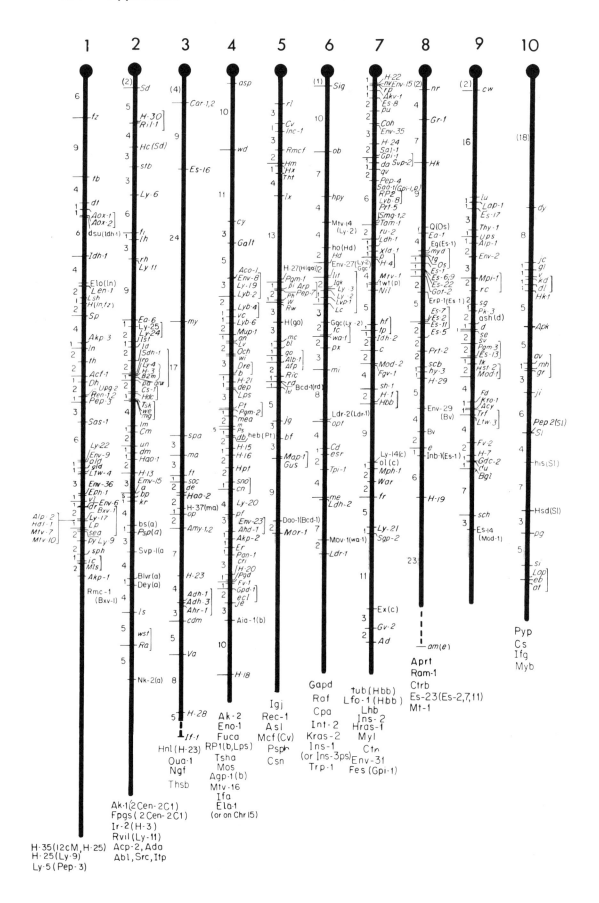

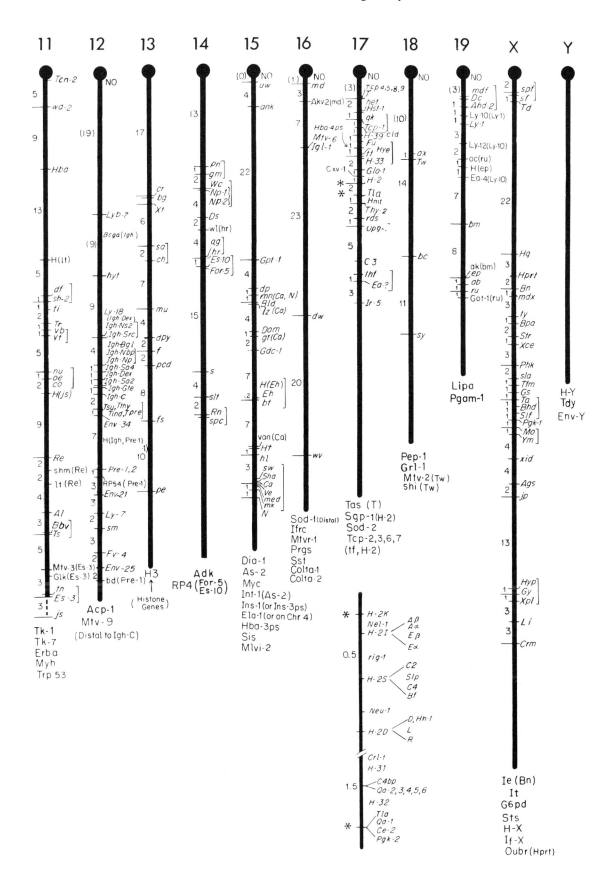

locus with which they have been located follows their symbol in parentheses. Straight brackets enclose closely linked loci whose order is not known with respect to each other. Curved brackets indicate uncertainty as to whether the genes are distinct loci. When the only information available is that a gene is "near" another gene, the nonitalicized symbol is placed at the bottom of the chromosome followed by the symbol of the locating locus in parentheses.

Acknowledgment

This work is supported by grant BSR 84-18828 from the National Science Foundation.

References

Davisson, M.T. and T.H. Roderick. 1981. Recombination percentages (ed. M.C. Green), pp. 283–313. *Genetic variants and strains of the laboratory mouse.* Gustav Fischer Verlag, Stuttgart.

Haldane, J.B.S., A.D. Sprunt, and N.M. Haldane. 1915. Reduplication in mice. *J. Genet. 5:* 133–135.

GENETIC LOCI OF THE MOUSE

Explanation: Genetic loci include both genes and DNA polymorphisms.
Recessive gene symbols begin with a lower case letter. Dominant gene symbols begin with an upper case letter. Semi-dominant gene symbols begin with an upper case letter but the gene name is followed by "SD".
Symbols followed by * indicate viral genes unique to certain strains.
Rules for nomenclature of DNA polymorphisms can be found in Mouse News Letter No. 72, Feb. 1975. UN means the chromosomal location of the locus is unknown. RE means this gene symbol is reserved for a locus yet to be identified in the mouse.

SYMBOL	NAME OF LOCUS	CHROMOSOME
a	non-agouti	2
Aal	active avoidance learning	UN
Aap	active avoidance performance	UN
ab	asebia	19
Abl	Abelson murine leukemia virus oncogene	2
Abp	androgen-binding protein (formerly Tcp)	17
Acc	anterior capsular cataract (provisional)	UN
Acf-1	albumin conformation factor-1	1
Ack	anti-C.kutscheri (formerly ack; ack=Ack[sl])	UN
Aco-1	aconitase-1	4
Aco-2	aconitase-2 mitochondrial, monomorphic (provisional)	UN
Acp-1	acid phosphatase-1	12
Acp-2	acid phosphatase-2	2
Acry-1	alpha-crystallin	17
Acy-1	aminoacylase-1	9
Ad	adult obesity and diabetes	7
Ada	adenosine deaminase	2
Adh-1	alcohol dehydrogenase-1	3
Adh-1t	alcohol dehydrogenase-1, temporal	3
Adh-3e	alcohol dehydrogenase-3, electrophoretic	3
Adh-3t	alcohol dehydrogenase-3, temporal	14
Adk	adenosine kinase	UN
adr	arrested development of righting response (provisional)	UN
Aem-1	antierythrocyte autoantibody (AEA) modifying gene (prov.)	4?
Afp	alpha fetoprotein	5
ag	agitans	14
Aglu	alpha-glucosidase	RE
Agp-1	alpha l-acid glycoprotein-1	4
Ags	alpha-galactosidase	X
Ah	aromatic hydrocarbon responsiveness (formerly In, Ahh)	17?
aha	autoimmune hemolytic anemia (provisional)	UN
Ahd-1	aldehyde dehydrogenase-1	4
Ahd-2	aldehyde dehydrogenase-2, cytoplasmic	19
Ahd-3	aldehyde dehydrogenase-3, microsomal	UN
Ahd-4	aldehyde dehydrogenase-4 (7 ± 2% recombination with Ahd-6)	UN
Ahd-5	aldehyde dehydrogenase-5	UN
Ahd-6	aldehyde dehydrogenase-6 (7 ± 2% recombination with Ahd-4)	UN
Ahd-3r	aldehyde dehydrogenase-3,regulator	UN
Ahr-1	aldehyde reductase-1	3
Aia-1	autoimmune hemolytic anemia-1	UN
Aia-2	autoimmune hemolytic anemia-2	19
ak	aphakia	19
Ak-1	adenylate kinase-1	2
Ak-2	adenylate kinase-2	4
Akp-1	alkaline phosphatase-1	1
Akp-2	alkaline phosphatase-2	1
Akp-3	alkaline phosphatase-3	7
Akv-1*	AKR leukemia virus inducer-1 (=Emv-11)	7
Akv-2*	AKR leukemia virus inducer-2 (=Emv-12)	16
Akv-3*	AKR leukemia virus inducer-3 (=Emv-13)	UN
Akv-4*	AKR leukemia virus inducer-4 (=Emv-14)	UN
Akvp*	AKR leukemia virus protein inducer (may = Emv-13)	UN
Akvr-1	AKR virus restriction locus-1 (= Fv-4)	UN
Al	alopecia	11
Ala-1	activated lymphocyte antigen-1 (=Ly-6)	UN
Alb-1	serum albumin variant	5
ald	adrenocortical lipid depletion	1
Aldo-1	aldolase-1	RE
Alm	anterior lenticonus with microphthalmia (provisional)	9
Alp-1	apolipoprotein A-I (formerly Sep-1)	1
Alp-2	apolipoprotein A-II	8
am	amputated	3
Amy-1	amylase-1	3
Amy-2	amylase-2	3
an	Hertwig's anemia	4
Anc	anisocoria (radiation induced eye mutation)(provisional)	UN
Anf	atrial natriuretic factor	UN
ank	progressive ankylosis	15
anx	anorexia	2
ao	apampischo	UN
Aox-1	aldehyde oxidase-1	1
Aox-2	aldehyde oxidase-2	1
ap	alopecia periodica	10
Apk	acid phosphatase-kidney (formerly Acp-2)	UN
Apo	anterior polar opacity (provisional)	UN
Apoc	anterior polar cataract (provisional)	UN
Aprt	adenine phosphoribosyl transferase	8
Apyc	anterior pyramidal cataract (provisional)	UN
ArP	Arp lymphoproliferative disease inducer	5
As-1	arylsulfatase A (formerly Arsa)	15
As-2	arylsulfatase B regulation	UN
asd	adipose storgae deficiency (provisional)	9
ash	ashen	9
Asl	argininosuccinate lyase	5
asp	audiogenic seizure prone	4
As-1r	arylsulfatase B regulation	UN
As-1s	arylsulfatase, temporal (provisional)	UN
at	atrichosis	10
Av-1	Abelson virus susceptibility-1	10
Av-2	Abelson virus susceptibility-2	UN
ax	ataxia	UN
azh	abnormal spermatazoon head	18
b	brown	4
B2m	beta-2 microglobulin	UN
ba	bare	UN
Badm	B-adrenergic binding (provisional)	11
Bbv*	B10.BR/SgLi endogenous ecotropic virus (B-tropic)	18
bal	balding	1
bc	bouncy	RE
Bcd-1	butyryl CoA dehydrogenase	12
Bcg	resistance to mycobacterium bovis BCG (probably = Lsh,Ity)	UN
Bcgf	B cell growth factor	5
Bdv-1*	BALB/c defective provirus-1	17
bf	buff	4?
Bf	complement component factor B	13
Bfo	bell-flash ovulation	UN
bg	beige	9
Bgb	beta-D-galactosidase in brain (provisional)	9
Bgl-e	beta-galactosidase, structural	9
Bgl-s	beta-galactosidase, systemic	9
Bgl-t	beta-galactosidase, temporal	9
bh	brain-hernia	7?
Bhd	broad headed	16

Symbol	Name	Chr
bl	blebbed	5
Bld	blind	15
bls	blanc-sale (provisional)	UN
Blvr	biliverdin reductase	19
bm	brachymorphic	X
Bn	bent-tail	X
bp	brachypodism	2
Bpa	bare-patches	X
Br	brachyrrhine (SD)	UN
Brp	brachypod Japan (provisional)	2
Bru	bruised	UN
Bsk	bare skin	UN
Bsp	black spleen	UN
Bst	blind-sterile	2
bt	belly spot and tail	16
bt-2	belted	15
bv	Bronx waltzer	9
Bv*	C57BL/10 endogenous ecotropic virus (N-tropic)	8
Bxv-1*	B10 xenotropic virus-1	1
c	albino	7
C2	complement component-2	17
C3	complement component-3 (formerly Plp)	17
C4	complement component-4 (formerly Ss)	17
C5	complement component-5 (see Hc)	2
C6	complement component-6	UN
C58v-1	C58 endogenous ecotropic virus-1	8
C58v-2	C58 endogenous ecotropic virus-2	UN
cab	caracul	15
cac	cardiac abnormality	UN
cad	recessive cataract	UN
Cad	congenital cataract (provisional)	UN
Car-1	carbonic anhydrase-1 (formerly Pre-1)	3
Car-2	carbonic anhydrase-2	3
Car-3	carbonic anhydrase-3 (muscle spec., monomorphic,provisional)	UN
Cat[AA]	cataract, Ann Arbor (provisional)	UN
Cat	dominant cataract	UN
cb	cerebral degeneration	UN
cby	chubby (provisional)	UN
Ccd	cleidocranial dysplasia (provisional)	6
Cd	crooked	RE
cda	cytidine deaminase	3
cdm	cadmium resistance	UN
Ce-1	liver catalase-1	17
Ce-2	kidney catalase-1	X
cg	circumgyrator (provisional)	UN
Cg	controller gene (provisional)	13
ch	congenital hydrocephalus	UN
cho	chondrodysplasia	UN
cht	chocolate (provisional)	UN
clf	clubfoot	1
cld	combined lipase deficiency	17
cls	coatloss (provisional)	2
Cm	coloboma	UN
Cms	cartilage matrix deficiency	UN
cmd	resistance to Coccidioides immitis	4
cn	achondroplasia	11
co	cocked	7
cod	cerebellar outflow degeneration	16
Coh	coumarin hydroxylase	16
Col1a-1	procollagen type I, alpha 1	RE
Col1a-2	procollagen type I, alpha 2	6
Cp	ceruloplasm	UN
Cpa	carboxypeptidase A	8
cpk	congenital polycystic kidneys (formerly ck)	UN
Cpl-1	plasma corticosterone level-1	UN
Cpl-2	plasma corticosterone level-2	UN
Cpz	chlorpromazine avoidance	13
cr	crinkled	UN
cri	cribiform degeneration	4
Crl-1	complement receptor lymphocyte frequency-1	17?
Crm	cream	X
Crz	control of early proliferation of T. cruzi (provisional)	UN
Cs	citrate synthase (formerly Cis, Cts)	10
Cs-1	catalase-1	2
Csn	casein gene family, -alpha, -beta, -gamma	5
ct	curly-tail	UN
Ctn	calcitonin	7
Ctrb	chymotrypsinogen B	8
Cts	cataract and small eye (SD) (formerly Cs)(provisional)	UN
Cv*	BALB/c ecotropic virus inducibility (=Emv-1)	5
cw	curly whiskers	9
Cxv-1	C xenotropic virus-1	17
Cxv-2	C xenotropic virus-2 (formerly Xensca)	4
Cyl	crinkly-tail lethal (provisional)	2
d	dilute (=Emv-3)	9
da	dark	7
Dac	dactylaplasia	UN
Dao	D-amino acid oxidase	5
db	diabetes	4
Dbr	dominant brown (provisional)	UN
Dbv	DBA/2J provirus (=Emv-3, d,)	9
Dc	dancer	19
Dce	desmosterol-to-cholesterol enzyme	RE
Dcp	dexamethasone-induced cleft palate (provisional)	17
de	droopy ear	3
dep	depilated	4
Dey	Dickie's small eye	2
df	Ames dwarf	11
Dfp	dark foot pads (SD)	UN
Dh	dominant hemimelia	1
Dhfr	dihydrofolate reductase	UN
di	duplicate incisors	UN
Dia-1	diaphorase-1 (NADH)	15
Dia-2	diaphorase-2 (NADPH)	RE
dl	downless	10
dm	diminutive	2
Dmm	disproportionate micromelia	UN
dn	deafness	UN
Dom	dominant megacolon	15
dp	dilution-Peru	15
Dpgm	diphosphoglyceromutase	RE
dpy	dumpy	13
dr	dreher	1
Dre	dominant reduced ear	4
Ds	disorganization	14
dsu	dilute suppressor	1
dt	dystonia musculorum	1
du	ducky	9
dup	dumpy-Oda (provisional)	UN
dv	dervish	UN
dw	dwarf	16
dy	dystrophia muscularis	10
dyl	dysgenetic lens	UN
e	recessive yellow (extension locus)	8
Ea-1	erythrocyte antigen-1	8
Ea-2	erythrocyte antigen-2 (formerly rho, R-Z, H-14)	UN

Symbol	Description	Chr
Ea-3	erythrocyte antigen-3 (formerly lambda)	UN
Ea-4	erythrocyte antigen-4 (formerly D)	UN
Ea-5	erythrocyte antigen-5 (formerly H-5)	UN
Ea-6	erythrocyte antigen-6 (formerly H-6)	2
Ea-7	erythrocyte antigen-7 (formerly T)	UN
Ea-8	erythrocyte antigen-8	UN
Ea-?	erythrocyte antigen-? (formerly Ea-2)	17
Eam	ethanol activity modifier	4?
eb	eye-blebs	10
ec	ectopic	UN
ecl	epistatic circling C57L/J	4
ecs	epistatic circling SWR/J	UN
Ed	extra digit	UN
Eg	endoplasmic glucuronidase	8
Eh	hairy ears	15
El	epilepsy (formerly ep)	UN
Ela-1	elastase-1 (may be pseudogene)	4
Ela-1	elastase-1 (may be pseudogene)	15
Elo	eye lens obsolescence	1
Ema	electrophoetic mobility and agglutinibility	UN
Emv-1	endogenous ecotropic murine leukemia virus-1(=Cv)	5
Emv-2	endogenous ecotropic murine leukemia virus-2(=Bv)	8
Emv-3	endogenous ecotropic murine leukemia virus-3(=d,Dbv,Sev-1)	9
Emv-4	endogenous ecotropic murine leukemia virus-4	UN
Emv-5	endogenous ecotropic murine leukemia virus-5	UN
Emv-6	endogenous ecotropic murine leukemia virus-6	UN
Emv-7	endogenous ecotropic murine leukemia virus-7	UN
Emv-8	endogenous ecotropic murine leukemia virus-8	UN
Emv-9	endogenous ecotropic murine leukemia virus-9	UN
Emv-10	endogenous ecotropic murine leukemia virus-10	UN
Emv-11	endogenous ecotropic murine leukemia virus-11(=Akv-1)	7
Emv-12	endogenous ecotropic murine leukemia virus-12(=Akv-2)	16
Emv-13	endogenous ecotropic murine leukemia virus-13(=Akv-3)	UN
Emv-14	endogenous ecotropic murine leukemia virus-14(=Akv-4)	UN
Emv-15	endogenous ecotropic murine leukemia virus-15(A[y]-assoc.)	2
Enc	embryonic nucleus cataract (provisional)	4
Eno-1	enolase-1	9
Env-2	envelope-2	UN
Env-5	envelope-5	UN
Env-6	envelope-6	1
Env-8	envelope-8	8?
Env-9	envelope-9	8?
Env-15	envelope-15	7
Env-21	envelope-21	UN
Env-23	envelope-23	UN
Env-25	envelope-25	UN
Env-27	envelope-27	6
Env-29	envelope-29	8
Env-31	envelope-31	7
Env-34	envelope-34	12
Env-35	envelope-35	7
Env-36	envelope-36	Y
Env-Y	envelope-Y	Y
Eo	eye-opacity	UN
Epa-1	pale ear	19
epf	epidermal antigen	UN
Eph-1	epileptiform (provisional)	UN
Er	epoxide hydrase	1
Erba	repeated epilation	4
Erp-1	avian erythroblastosis virus oncogene a	11
Es-1	esterase-1	8
Es-2	esterase-2	8
Es-3	esterase-3	11
Es-5	esterase-5	8
Es-6	esterase-6	8
Es-7	esterase-7	8
Es-8	esterase-8	7
Es-9	esterase-9	8
Es-10	esterase-10	14
Es-11	esterase-11	8
Es-12	esterase-12	UN
Es-13	esterase-13	9
Es-14	esterase-14	9
Es-15	esterase-15	UN
Es-16	esterase-16	3
Es-17	esterase-17	9
Es-18	esterase-18	UN
Es-19	esterase-19	8
Es-20	esterase-20 (prob. = a trimer of Es-9 and Es-6)	UN
Es-21	esterase-21 (may = Es-13)	8
Es-22	esterase-22	8
Es-23	esterase-23	8
Es-23t	esterase-23, temporal	8
Es-24	esterase-24	UN
Es-25	esterase-25	6
Esr	esterase-5 regulator	7
Exa	earlier X-zone degeneration	4?
Ex	exploratory activity	UN
exf	exfoliate (formerly ex)	9
Exnm	neuromuscular excitability	UN
ey-1	eyeless-1	UN
ey-2	eyeless-2	UN
Ezg	extent of zona glomerulosa	13
f	flexed-tail	UN
fa	falter	UN
far	first arch defect	UN
fat	fat	UN
Fbp-1	fructose bisphosphatase	2
fc	flecking	9
fd	fur deficient	7
Fes	feline sarcoma virus oncogene	7
Fgv-1*	C3H/Fg virus-1	UN
Fgv-2*	C3H/Fg virus-2	UN
fh-1	fetal hematoma	RE
Fhe	Friend helper virus erythroblastosis susceptibility	UN
fi	fidget	2
Fk	fleck (SD)	UN
fl	flipper-arm	UN
Flv	flavivirus resistance	UN
fm	foam-cell reticulosis	UN
Fms	feline sarcoma virus virus oncogene, McDonough strain	UN
fmt	formentin	UN
For-1	formidase-1	UN
For-2	formidase-2	UN
For-3	formidase-3	UN
For-4	formidase-4	14
For-5	formidase-5	12
Fos	FBJ murine osteosarcoma virus oncogene	2
Fpgs	folypolyglutamyl synthetase	7
fr	frizzy	UN
fri	frissonant (provisional)	UN
Frl	furloss (SD) (provisional)	UN
fro	fragilitas ossium	13
fs	furless	3
ft	flaky tail	3
Fu	fused	17
Fuca	alpha-fucosidase	4
Fv-1	Friend virus susceptibility-1	4

Symbol	Description	Chr
Fv-2	Friend virus susceptibility-2	9
Fv-3	Friend virus susceptibility-3	UN
Fv-4	Friend virus susceptibility-4 (=Akvr-1)	12
Fv-5	Friend virus susceptibility-5	UN
Fw	Friend virus P-induced early anemia, polycythemia susc.	5?
fz	fawn (provisional)	UN
fzt	fuzzy	1
fzt	fuzzy tail	UN
Ga	greying with age	UN
Gad-1	glutamic acid decarboxylase-1 (provisional)	UN
Gad-2	glutamic acid decarboxylase-2 (provisional)	UN
Galt	galactose-1-phosphate uridyl transferase	4
Gapd	glyceraldehyde-3-phosphate dehydrogenase	6
Gaps	phosphoribosyl glycinamide synthetase	RE
Gde	guanidine deaminase-morphic	UN
Gdc-1	L-glycerol 3-phosphate dehydrogenase	15
Gdc-1	glycerol-3-phosphate dehydrogenase	9
Gdct-1	GPDH mRNA regulation-1	UN
Gdr-1	GPDH mRNA regulation-2	UN
Gdr-2	G6PD regulator-1	UN
Ggc	G6PD regulator-2	UN
Gk	gamma-glutamyl cyclotransferase	6
Gk	glucokinase	UN
gl	grey-lethal	10
gld	generalized lymphoproliferative disease	1
Glk	galactokinase	11
Glo-1	glyoxylase-1	17
gm	gunmetal	14
go	angora	5
Got-1	glutamate oxaloacetate transaminase-1	19
Got-2	glutamate oxaloacetate transaminase-2	8
gp	gaping lids	UN
Gpd-1	glucose-6-phosphate dehydrogenase-1	X
Gpdx	glucose-6-phosphate dehydrogenase-1, X-linked (form. G6pd)	X
Gpi-1s	glucose phosphate isomerase-1, structural	7
Gpi-1t	glucose phosphate isomerase-1, temporal	7
Gpt-1	glutamic-pyruvic transaminase-1, soluble	15
Gpx	glutathione peroxidase	RE
gr	grizzled	10
Grl-1	glucocorticoid receptor-1	18
gs	glabrous (provisional)	UN
gs	greasy	X
Gss	glutamate-gamma-semialdehyde synthetase	X
gt	gray tremor	RE
Guk-1	guanylate kinase-1	15
Gus-r	beta-glucuronidase, regulator	RE
Gus-s	beta-glucuronidase, structural	5
Gus-t	beta-glucuronidase, temporal	5
Gus-u	beta-glucuronidase-systemic regulator	5
Gv-1*	Gross virus antigen-1	UN
Gv-2*	Gross virus antigen-2	7
Gy	gyro	X

LIST OF SOME LOCI OR REGIONS CONTAINED WITHIN THE H-2 COMPLEX:

Symbol	Description	Chr
H-1	histocompatibility-1	7
H-2	histocompatibility-2 (major histocompatibility complex)	17
H-2D	(controls antigenic specificities of D-region loci)	17
H-2G	(serologically detected antigen abundant on erythrocytes)	17
H-2I	(immune response region)	17
H-2K	(antigens for graft-versus-host [GVH] mixed lymphocyte [MLR] and cell mediated lysis [CML] reactions)	17
H-2L	(controls antigenic specificities H-2.1, H-2.28)	17
H-2S	(region contains Ss and Slp)	17
H-2T	(target antigens reactive in the cell-mediated lympholysis assay)	17
H-3	histocompatibility-3	2
H-4	histocompatibility-4	7
H-7	histocompatibility-7	9
H-8	histocompatibility-8	UN
H-9	histocompatibility-9	UN
H-10	histocompatibility-10	UN
H-11	histocompatibility-11	UN
H-12	histocompatibility-12	2
H-13	histocompatibility-13	2
H-15	histocompatibility-15	4
H-16	histocompatibility-16	4
H-17	histocompatibility-17	UN
H-18	histocompatibility-18	4
H-19	histocompatibility-19	8
H-20	histocompatibility-20	4
H-21	histocompatibility-21	7
H-22	histocompatibility-22	3
H-23	histocompatibility-23	7
H-24	histocompatibility-24	1
H-25	histocompatibility-25	UN
H-26	histocompatibility-26	5
H-27	histocompatibility-27	3
H-28	histocompatibility-28	3
H-29	histocompatibility-29	8
H-30	histocompatibility-30	2
H-31	histocompatibility-31	17
H-32	histocompatibility-32	17
H-33	histocompatibility-33	17
H-34	histocompatibility-34	UN
H-35	histocompatibility-35	1
H-36	histocompatibility-36	UN
H-37	histocompatibility-37	UN
H-38	histocompatibility-38	UN
H-39	histocompatibility-39	17
H-41	histocompatibility-41	UN
H(Eh)	histocompatibility(Eh)(provisional)	15
H(ep)	histocompatibility(ep)(provisional)	19
H(go)	histocompatibility(go)(provisional)	5
H(igh)	histocompatibility(igh)(provisional) (prob. = H-40)	12
H(js)	histocompatibility(js)(provisional)	11
H(ln,fz)	histocompatibility(ln,fz)(provisional)	1
H(lt)	histocompatibility(lt)(provisional)	11
H(pi)	histocompatibility(pi)(provisional)	5
H(tn)	histocompatibility(tn)(provisional)	11
H-X	histocompatibility-X	X
H-Y	histocompatibility-Y	Y
H3	histone gene complex (provisional)	13
Hadh	hydroxyacyl-CoA-dehydrogenase	RE
Hao-1	hydroxyacid oxidase-1, liver	2
Hao-2	hydroxyacid oxidase-2, kidney	3
Hat-1	histidine amino-transferase-1	UN
Hba	hemoglobin alpha-chain	11
Hba-3ps	hemoglobin alpha-3, pseudogene	15
Hba-4ps	hemoglobin alpha-4, pseudogene	17
Hbb	hemoglobin beta-chain	7
hbd	hemoglobin deficient	UN
Hbx	hemoglobin alpha-like embryonic chain ((in Hba complex)	11
Hby	hemoglobin beta-like embryonic chain (in Hbb complex)	7
Hc	hemolytic complement, (= C5)	2
hcp	hypochondrodysplasia (provisional)	UN
Hct	hair constriction	UN
Hd	hypodactyly	6
Hdc-c	histidine decarboxylase concentration, kidney	2

Symbol	Description	Chr
Hdc-e	histidine decarboxylase estrogen response	2
Hdc-s	histidine decarboxylase, structural, kidney	2
Hdl-1	high density lipoprotein-1	1
hea	hereditary erythroblastic anaemia	UN
heb	head blebs	4
Hemb	hemophilia B	RE
het	head-tilt	17
Hex-1	hexoseaminidase-1	RE
hf	hepatic fusion	7
Hh-1	hemopoietic histocompatibility (within H-2D)	17
Hi	hare tail (provisional)	UN
hid	hair interior defect	UN
his	histidinemia	10
Hk	hook	8
Hk-1	hexokinase-1	10
hl	hair-loss	15
Hld	hippocampal lamination defect	UN
Hm	hammer-toe	5
Hma	HPRT mobility alteration	7
Hmt	histocompatibility maternally transmitted	17
Hnl	hypothalamic norepinephrine level (near H-23)	3
ho	hotfoot	6
Hom-1	hormone metabolism-1 (possibly within H-2)	17?
hop	hop-sterile	UN
hpc	hyperspiny Purkinge cell (provisional)	UN
hpg	hypogonadal	UN
Hprt	hypoxanthine-guanine phosphoribosyl transferase	X
Hpt	hair patches	4
hq	harlequin	X
hr	hairless	14
Hras-1	Harvey rat sarcoma virus oncogene	7
Hrt-1	heart protein-1	UN
Hsd	histidase synthetic rate	10
Hsp70	heat shock protein 70 (provisional)	UN
Hst-1	hybrid sterility-1	17
Hst-2	hybrid sterility-2	9
Hsv-1	herpes simplex virus-1 (provisional, = If-X)	X
Ht	hightail	15
Hti	histidine decarboxylase thyroxin inducibility	UN
Hv	hepatitis virus susceptibility	UN
Hx	hemimelic extra toes	5
hy-3	hydrocephalus-3	8
Hye	H-Y expression	17
Hyp	hypophosphatemia	X
hyt	hypothyroidism	12
Ia-1	I-region-associated-antigen-1	17
Ia-2	I-region-associated-antigen-2	17
Ia-3	I-region associated antigens	17
Ia-4	I-region-associated-antigen-4	17
Ia-5	I-region-associated-antigen-5	17
Iac	iris anomaly with cataract (provisional)	UN
Ias	inhibition of audiogenic seizures (linked to Ah)	17?
ic	ichthyosis	1
Idc	Iris dysplasia with cataract (provisional)	UN
Idh-1	isocitrate dehydrogenase-1	1
Idh-2	isocitrate dehydrogenase-2	7
Idh-3	isocitrate dehydrogenase-3	UN
Ie	eye-induced reduction	X
If-1	MTV-induced circulating interferon	3
If-2	MuLV-induced circulating interferon	UN
If-3	Sendai virus-induced circulating interferon-1	UN
If-4	Sendai virus-induced circulating interferon-2	UN
If-X	NDV-induced circulating interferon, x-linked	X
Ifa	interferon, alpha	4
Iff	interferon, fibroblast	RE

Symbol	Description	Chr
Ifg	interferon, gamma	10
Ifl	interferon, leukocyte	RE
Ifrc	interferon receptor	16
Igh	immunoglobulin heavy chain complex	12
Igh-C	immunoglobulin heavy-chain constant region	12

LIST OF LOCI WITHIN THE Igh-C REGION WHICH HAVE BEEN SHOWN BY MOLECULAR METHODS TO BE NONALLELIC

Symbol	Description	Chr
Igh-1	immunoglobulin heavy chain-1 (serum IgG2a)	12
Igh-2	immunoglobulin heavy chain-2 (serum IgA)	12
Igh-3	immunoglobulin heavy chain-3 (serum IgG2b)	12
Igh-4	immunoglobulin heavy chain-4 (serum IgG1)	12
Igh-5	immunoglobulin heavy chain-5 (delta-like heavy chain of an IgD-like cell surface immunoglobulin)	12
Igh-6	immunoglobulin heavy chain-6 (heavy chain of IgM)	12
Igh-V	immunoglobulin heavy-chain variable region	12

LIST OF LOCI WITHIN THE Igh-V REGION WHICH HAVE BEEN SHOWN BY RECOMBINATION TO BE NONALLELIC

Symbol	Description	Chr
Igh-Bgl		12
Igh-Dex		12
Igh-Gte		12
Igh-Nbp		12
Igh-Np		12
Igh-Ns2		12
Igh-Sa2		12
Igh-Sa4		12
Igh-Src		12

LIST OF LOCI MAPPED TO THE Igh-V REGION BUT NOT PROVED TO BE NONALLELIC

Symbol	Description	Chr
Igh-Aal		12
Igh-Aa2		12
Igh-Aa3		12
Igh-Ars		12
Igh-Inu		12
Igh-Lev		12
Igh-Ns1		12
Igh-Ns2		12
Igh-Ns3		12
Igh-Ns4		12
Igh-Ns5		12
Igh-Pc		12
Igh-Sa1		12
Igh-Sa3		12
Igh-Sa5		12

Symbol	Description	Chr
Igj	immunoglobulin joining gene (not J region)	5
Igk	immunoglobulin kappa gene complex	6
Igk-C	immunoglobulin kappa chain constant region	6
Igk-V	immunoglobulin kappa chain variable region	6

LIST OF LOCI WITHIN THE Igk-V REGION WHICH HAVE BEEN SHOWN TO BE NONALLELIC

Symbol	Description	Chr
Igk-Efl		6
Igk-Pc		6
Igk-Trp		6

Symbol	Description	Chr
Igl-1	immunoglobulin lambda-1 chain	16
Igl-1r	immunoglobulin lambda-1 chain regulator	16
Ihe-1	intestinal helminth expulsion-1 (provisional)	UN
Il-1	interleukin-1	RE
Il-2	interleukin-2	UN
Inb-1	inducibility of MuLV in C57BL	UN
Inc-1	inducibility of MuLV in BALB/c	UN
Ins-1	insulin I or pseudogene to insulin I (or on Chr 15)	6

Symbol	Description	Chr
Ins-3ps	Insulin I or pseudogene to insulin I (or on Chr 6)	15
Ins-2	Insulin II	7
Int	temporal control of H-2 expression (maps close to H-2 cis control; see Rec)	17
int-1	mammary tumor integration site - 1	15
int-2	mammary tumor integration site - 2	7
Ipo-1	indophenol oxidase-1	UN
Irr-1	immune response-1 (within H-2I region)	17
Ir-2	immune response-2	2
Ir-3	immune response-3	UN
Ir-4	immune response-4	UN
Ir-5	immune response-5	17
Ir-7	immune response-7	UN
It	irregular teeth	X
Itp	inosine triosephosphate	2
Ity	immunity to S. typhimurium (=Lsh,Bcg probably)	1
iv	situs inversus viscerum	UN
ja	jaundiced	UN
jc	Jackson circler	10
je	jerker	4
jg	jagged-tail	5
ji	jittery	10
jp	jimpy	X
jpk	juvenile polycystic kidneys	UN
js	Jackson shaker	11
jsd	juvenile spermatogonial depletion	UN
jt	joined toes	UN
jv	Jackson waltzer	UN
Kap	kidney androgen-regulated protein	UN
kd	kidney disease	10
Kfo-1	kidney 40-50 thousand M.Wt. protein-1 (provisional)	9
kr	kreisler	2
Kras-1	Kirsten rat sarcoma virus oncogene-1, pseudogene	RE
Kras-2	Kirsten rat sarcoma virus oncogene-2, expressed	6
Kth-1	kidney 30-40 thousand M.Wt. protein-1 (provisional)	UN
Kth-2	kidney 20-40 thousand M.Wt. protein-2 (provisional)	UN
Kw	kinky-waltzer	UN
ky	kyphoscoliosis	UN
Lad	lymphocyte-activity determinants	17?
Lap-1	leucine arylaminopeptidase-1, intestinal	9
Lap-2	leucine arylaminopeptidase-2, serum (provisional)	UN
Lbt-1	lupus band test-1	17
Lc	lurcher	6
Lcam	liver cell adhesion molecule	RE
Lcat	lecithin cholesterol acyltransferase	RE
ld	limb deformity	2
Ldh-1	lactate dehydrogenase-1	7
Ldh-2	lactate dehydrogenase-2	6
Ldr-1	lactate dehydrogenase regulator-1	6
Ldr-2	lactate dehydrogenase regulator-2	6
le	light ear	5
lec	laryngotracheo-esophageal cleft (provisional)	7
Len-1	eye lens protein-1 (provisional)	1
Len-2	eye lens protein-2	UN
Lfo-1	liver 40-50 thousand M.Wt. protein-1 (provisional)	7
Lfo-2	liver 40-50 thousand M.Wt. protein-2 (provisional)	UN
lg	lid gap	UN
lgr	London grey	UN
lh	lethargic	2
Lhb	luteinizing hormone beta subunit	7
Li	lined	X
Lip-1	lysosomal acid lipase-1 (formerly Lipa)	19
lit	little	6
lm	lethal milk	2
ln	leaden	1
Lna-1	lymph node antige-1	UN
Lop	lens opacity	10
Lop-2	lens opacity-2 (provisional)	UN
Lop-3	lens opacity-3 (provisional)	UN
Lop-4	lens opacity-4 (provisional)	2?
Lop-5-9	lens opacity-5 through -9 (provisional)	UN
Lp	loop-tail	1
Lpn-1	NZ lupus nephritis-1 (formerly Ln-1)	17
Lpn-2	NZ lupus nephritis-2 (formerly Ln-2)	17
Lpn-3	NZ lupus nephritis-3	UN
lpr	lymphoproliferation	4
Lps	lipopolysaccharide response	UN
lr	lens rupture	2
ls	lethal spotting	1
Lsh	Leishmaniasis resistance (probably = Ity, Bcg)	UN
lst	Strong's luxoid	2
lt	lustrous	1
Lth-2	liver 30-40 thousand M.Wt. protein-2 (provisional)	11
Ltn-2	liver 10-20 thousand M.Wt. protein-2 (provisional)	1
Ltw-2	liver 20-30 thousand M.Wt. protein-3 (provisional)	12
Ltw-3	liver 20-30 thousand M.Wt. protein-4 (provisional)	9
Ltw-4	liver 20-30 thousand M.Wt. protein-5 (provisional)	1
Ltw-6	liver 20-30 thousand M.Wt. protein-6 (provisional)	9
lu	luxoid	UN
Lus	lymphoid cytostasis suppressor	9
Lv	delta-aminolevulinate dehydratase	4
Lvp-1	major liver protein-1	5
lx	luxate	6
Ly-1	lymphocyte antigen-1 (formerly Lyt-1)	19
Ly-2	lymphocyte antigen-2 (formerly Lyt-2)	6
Ly-3	lymphocyte antigen-3 (formerly Lyt-3)	6
Ly-4	lymphocyte antigen-4	2
Ly-5	lymphocyte antigen-5	2
Ly-6	lymphocyte antigen-6 (formerly Ly-8,Ala,DAG,H9/25)	9
Ly-7	lymphocyte antigen-7	12
Ly-8	lymphocyte antigen-8 (formerly Ly-11)	1
Ly-9	lymphocyte antigen-9 (formerly Lgp100,T100)	1
Ly-10	lymphocyte antigen-10	19
Ly-11	lymphocyte antigen-11 (formerly Ly-m11)	UN
Ly-12	lymphocyte antigen-12	UN
Ly-13	lymphocyte antigen-13	7
Ly-14	lymphocyte antigen-14	UN
Ly-15	lymphocyte antigen-15	12
Ly-16	lymphocyte antigen-16 (formerly Ly-18)	1
Ly-17	lymphocyte antigen-17 (= Ly-20, LyM-1, Ly-m1)	12
Ly-18	lymphocyte antigen-18 (formerly Ly-m18) linked to Ltw-2	12
Ly-19	lymphocyte antigen-19 (formerly Ly-m19)	4
Ly-20	lymphocyte antigen-20 (formerly Ly-22a)	7
Ly-21	lymphocyte antigen-21	2
Ly-22	lymphocyte antigen-22 (formerly Ly-m22)	2
Ly-23	lymphocyte antigen-23	2
Ly-24	lymphocyte antigen-24 (part of Ly-6 complex)	4
Ly-26	lymphocyte antigen-26 (formerly Pgp-1)	UN
Lyb-2	B-lymphocyte antigen-2	4
Lyb-3	B-lymphocyte antigen-3	UN
Lyb-4	B-lymphocyte antigen-4	4
Lyb-5	B-lymphocyte antigen-5	UN
Lyb-6	B-lymphocyte antigen-6	4
Lyb-7	B-lymphocyte antigen-7	12
Lyb-8	B-lymphocyte antigen-8	7
lz	lizard	15

Chr	Symbol	Name
4	m	misty
3	ma	matted
1	Mal-1	malaria resistance (may = Lsh, Ity)
5	Map-1	mannosidase processing-1
18	Mbp	myelin basic protein (= shi)
5	mc	marcel
UN	Mch	modifier of chinchilla (provisional)
UN	Mcm-1	modifier of c[mj]-1
UN	Mcm-2	modifier of c[mj]-2 (provisional)
16	md	mahoganoid
UN	mdac	dactylaplasia modifier (controls expression of Dac)
19	mdf	muscle deficient
UN	mdg	muscle dysgenesis
7	Mdr	malic enzyme mitochondrial-regulatory (near Mod-2)
X	mdx	X-linked muscular dystrophy (formerly pke)
6	me	motheaten
4	mea	meander tail
15	med	motor end-plate disease
2	ng	mahogany
10	mh	mocha
6	mi	microphthalmia
UN	mic	microphthalmia Japan (provisional)
15	mk	microcytic anemia
1	ml	myelin-less (provisional)
UN	Mls	minor MLC-stimulating
15	Mlvi-1*	murine leukemia virus-1
15	Mlvi-1	Moloney-MuLV integration site-1
15	Mlvi-2	Moloney-MuLV integration site-2
X	mn	miniature
9	Mo	mottled
7	Mod-1	malic enzyme, supernatant
5	Mod-2	malic enzyme, mitochondrial
RE	Mor-1	malate dehydrogenase, mitochondrial
4	Mor-2	malate dehydrogenase, soluble
6	Mos	Moloney sarcoma virus oncogene
UN	Mov-1	Moloney leukemia virus-1
7	Mp	micropinna-microphthalmia (SD)
9	Mph-1	macrophage antigen-1
UN	Mpi-1	mannosephosphate isomerase-1
8	Msp-1	mouse salivary protein-1 (provisional)
8	Mt-1	metallothionein-1
UN	Mt-2	metallothionein-2
UN	mto	myotonia
7	Mtp-2	mouse tear protein-2
18	Mtv-1*	mammary tumor virus locus-1
11	Mtv-2*	mammary tumor virus locus-2
UN	Mtv-3*	mammary tumor virus locus-3
UN	Mtv-4*	mammary tumor virus locus-4
16	Mtv-5*	mammary tumor virus locus-5
1	Mtv-6*	mammary tumor virus locus-6
UN	Mtv-7*	mammary tumor virus locus-7 (prob. = Mtv-10)
12	Mtv-8*	mammary tumor virus locus-8
UN	Mtv-9*	mammary tumor virus locus-9
UN	Mtv-10*	mammary tumor virus locus-10
UN	Mtv-11*	mammary tumor virus locus-11
UN	Mtv-12*	mammary tumor virus locus-12
6	Mtv-13*	mammary tumor virus locus-13
UN	Mtv-14	mammary tumor virus locus-14
4	Mtv-15*	mammary tumor virus locus-15
UN	Mtv-16*	mammary tumor virus locus-16
16	Mtv-18	mammary tumor virus locus-18
13	Mvr-1	mammary tumor virus receptor-1
4	mu	muted
UN	mut-1	mutation-1 (provisional)
UN	mut-2	mutation-2 (provisional)
UN	Mv	malformed vertebrae (SD)
5?	Mx	myxovirus (influenza virus) resistance
UN	Mxv-1	MA/My xenotropic MuLV-1
3	my	blebs
10	Myb	avain myeloblastosis virus oncogene
15	Myc	myelocytomatosis virus oncogene
8	myd	myodystrophy
11	Myh	myosin heavy chain, skeletal
7	Myl	myosin light chain, skeletal
15	N	naked
UN	Nat	liver N-acetyl transferase
UN	nb	normoblastic anemia
RE	Ncam	neural cell adhesion molecule
UN	nct	Nakano cataract
17	Nel-1	CML-detected lymphocyte antigen (provisional)
17	Neu-1	neuraminidase-1 (formerly Apl, Map-2, Aglp)
UN	Ng	nackig (SD)
RE	Ngcam	neural glial cell adhesion molecule
3	Ngf	nerve growth factor, beta subunit
7	Nil	neonatal intestinal lipidosis
2	Nk-1	NK-associated antigen-1
14	Nk-2	NK-associated antigen-2
14	Np-1	nucleoside phosphorylase-1
8	Np-2	nucleoside phosphorylase-2
11	nr	nervous
UN	nu	nude
7	nuc	nuclear cataract (provisional)
UN	Nuc	nuclear cataract (provisional, needs change?)
UN	nv	Nijmegen waltzer
X	Nxv-1	NZB/BlNJ xenotropic MuLV (provisional)
5	Nxv-2	NZB/BlNJ xenotropic MuLV (provisional)
6	Nzc	nuclear and zonular cataract (provisional)
19	Nzv-1*	NZB virus-1
4	Nzv-2*	NZB virus-2
UN	O[hv]	organizer, high variegation (provisional)
UN	O[slG]	operator noninducible (provisional)
9	ob	obese
19	oc	osteosclerotic
UN	Och	ochre
8	oe	open eyelids
UN	oed	edematous
UN	oel	open eyelids with cleft palate
7	oh	obstructive hydrocephalus
18	ol	oligodactyly
11	olt	oligotriche
UN	om	ovum mutant (SD)
16	op	osteopetrosis
1	opt	opisthotonus
UN	or	ocular retardation
12	Os	oligosyndactylism
UN	ot	oscillator
UN	Oua-1	ouabain resistance
UN	Ox-1	menadione oxidoreductase-1
UN	Ox-2	menadione oxidoreductase-2
7	p	pink-eyed dilution
2	pad	paddle (provisional)
UN	Pan-1	pancreas protein-1
4	par	paralyse (provisional)
16	pc	phocomelic
UN	Pca-1	plasma cell antigen-1
13	pcd	Purkinje cell degeneration
UN	pcp	polydactyly with cleft palate (provisional)

Symbol	Description	Chr
Pcs	polar cataract and small eye (provisional)	UN
Pct	plasmacytomagenesis	UN
Pd	pyrimidine degrading	UN
pdn	polydactyly Nagoya	UN
pe	pearl	13
Pep-1	peptidase-1	18
Pep-2	peptidase-2	10
Pep-3	peptidase-3	1
Pep-4	peptidase-4	7
Pep-7	peptidase-7	5
Pfk	phosphofructokinase	4
pf	pupoid fetus	RE
pg	pygmy	10
Pgam-1	phosphoglyceromutase-1	19
Pgd	6-phosphogluconate dehydrogenase	4
Pgk-1	phosphoglycerate kinase-1	X
Pgk-2	phosphoglycerate kinase-2	17
Pgm-1	phosphoglucomutase-1	5
Pgm-2	phosphoglucomutase-2	4
Pgm-3	phosphoglucomutase-3	9
Ph	patch	X
Phk	phosphorylase kinase	5
Phr	pheromonal response	UN
pi	pirouette	UN
Pim-1	proviral integration site MuLV-1 (provisional)	UN
Pk-1	pyruvate kinase-1	UN
Pk-2	pyruvate kinase-2	9
Pk-3	pyruvate kinase-3	UN
pma	peroneal muscular atrophy	UN
pn	pugnose	14
Pnc	pancreas defect	UN
Po	postaxial polydactyly	UN
Pomc-2	pro-opiomelanocortin-beta	19?
Pp	passive performance	4
pt	porcine tail	UN
Pre-1	prealbumin-1 (= alpha-1-antitrypsin)	12
Pre-2	prealbumin-2	12
Prgs	phoshoribosylglycinamide synthetase	UN
Pro-1	proline oxidase-1	UN
Prt-1	pancreatic proteinase-1 (no recomb. with Prt-3)	8
Prt-2	pancreatic proteinase-2	UN
Prt-3	pancreatic proteinase-3 (no recomb. with Prt-1)	UN
Ps	polysyndactyly	4
Psp	parotid secretory protein	2
Psph	phosphoserine phosphatase (formerly Psp)	5
Pt	pintail	4
Pth	parathyroid hormone	RE
ptr	pulmonary tumor resistance	UN
pu	pudgy	7
Pv	pivoter (probably SD)	UN
px	postaxial hemimelia	6
Py	polydactyly	1
Pyp	inorganic pyrophosphatase	10
Q	quinky	8
Qa-1	Qa lymphocyte antigen-1	17
Qa-2	Qa lymphocyte antigen-2	17
Qa-3	Qa lymphocyte antigen-3	17
Qa-4	Qa lymphocyte antigen-4 (formerly Qat-4)	17
Qa-5	Qa lymphocyte antigen-5 (formerly Qat-5)	17
Qa-6	Qa lymphocyte antigen-6 (provisional)	17
Qa-7	Qa lymphocyte antigen-7 (provisional)	17
Qa-8	Qa lymphocyte antigen-8 (provisional)	17
Qa-9	Qa lymphocyte antigen-9 (provisional)	17
Qed-1	Qed lymphocyte antigen-1 (may = Qa-1)	17

Symbol	Description	Chr
qk	quaking	17
Qui	quinine sensitivity	UN
qv	quivering	7
r	rodless retina	10?
Ra	ragged	2
Raf	alpha fetoprotein regulation, adult (provisional)	UN
Raf-1	ras-related fibrosarcoma virus oncogene-1	6
Raf-2	ras-related fibrosarcoma virus oncogene-2	8
Ram-1	replication of amphotropic virus-1	UN
ras	resistance to audiogenic seizures	9
rc	rough coat	5
Rcs-1	reticular cell sarcoma suppression-1 (provisional)	17
rds	retinal degeneration	11
Re	retinal degeneration, slow	UN
Rec	temporal control of H-2 expression (not H-2 linked, 20cM from Tem)	5
Rec-1	replication of ecotropic virus-1	1
Ren-1	renin (formerly Rnr[b],Rn-1,Ren-A)	1
Ren-2	renin (formerly Rnr[s],Rn-2,Ren-B)	1
Rf	rib fusion (SD)	17
Rfv-1	recovery from Friend virus-1	17
Rfv-2	recovery from Friend virus-2	17
Rfv-3	recovery from Friend virus-3	UN
rg	rotating	17
Rgv-1	resistance to Gross virus-1	17
Rgv-2	resistance to Gross virus-2	UN
rh	rachiterata	2
rhg	retarded hair growth	UN
Rhv-1	resistance to hepatitis virus-1 (see Hv-1)	UN
Rhv-2	resistance to hepatitis virus-2 (see Hv-1)	UN
Ri-1	recognition of identity-1	17
Ri-2	recognition of identity-2	5
Ric	rickettsia tsutsugamushi resistance	UN
rif	alpha fetoprotein regulation, inducibility	UN
Rig-1	regulation of Igh-1[b]-1	17
Rig-2	regulation of Igh-1[b]-2 (provisional)	12
Ril-1	radiation-induced leukemia sensitivity-1	2
Ril-2	radiation-induced leukemia sensitivity-2	4?
Ril-3	radiation-induced leukemia sensitivity-3	5
rl	reeler	17
Rmc-1	MCF virus sensitivity (putative MCF receptor locus)	17
Rmcf-1	MCF virus replication restriction	17
Rmv-1	resistance to Moloney virus-1 (provisional)	14
Rmv-2	resistance to Moloney virus-2 (provisional)	2
Rmv-3	resistance to Moloney virus-3 (provisional)	7
Rn	roan	4
ro	rough	7
rp	reduced pigmentation	7
RP1	Roswell Park 1, DNA polymorphism (provisional)	8
RP2	Roswell Park 2, DNA polymorphism (provisional)	14
RP2-s	Roswell Park 2, structural	UN
RP2-r	Roswell Park 2, regulator	UN
RP3	Roswell Park 3, DNA polymorphism (provisional)	17
RP4	Roswell Park 4, DNA polymorphism (provisional)	14
RP5	Roswell Park 5, DNA polymorphism (provisional)	UN
RP6	Roswell Park 6, DNA polymorphism (provisional)	UN
RP10	Roswell Park 10, DNA polymorphism (provisional)	17
RP11	Roswell Park 11, DNA polymorphism (provisional)	17
RP17	Roswell Park 17, DNA polymorphism (provisional)	17
RP54	Roswell Park 54, DNA polymorphism (provisional)	12
Rrs	resistance to Rous sarcoma	17
Rrv-1	resistance to RadLV-1	5
rs	recessive spotting (could be allele at w)	UN
rst	rosette	

Symbol	Description	Chr
Rtp	resistance to transplantable plasmacytoma MPC-11 (provis.)	UN
ru	ruby-eye	19
ru-2	ruby-eye-2	UN
Rvl-1	Rauscher leukemia virus susceptibility-1	UN
Rvil-1	radiation-induced leukemia virus susceptibility	1
Rw	rump white	5
s	piebald	14
sa	satin	13
Sac	saccharin preference	UN
Sal	satin-like (provisional)	UN
Salv-1	salivary protein variant-1 (provisional)	7
Sas-1	serum antigenic substance-1	1
scb	scabby	8
sch	scant hair	9
scid	severe combined immunodeficiency	UN
Scl	susceptibility to cutaneous Leishmaniasis	8
Sco	scopolamine modification of exploratory activity	17
Sd	Danforth's short tail	2
Sdh-1	sorbitol dehydrogenase-1	2
Sdr-1	serine dehydratase regulator-1	UN
Sdr-2	serine dehydratase regulator-2	UN
se	short ear	9
sea	sepia	1
Sey	small eye (SD)	2
sf	scurfy	X
sg	staggerer	9
Sgp-1	serum gp70 production-1 (provisional)	17
Sgp-2	serum gp 70 production-2 (provisional)(possibly Gv-2)	7
sh-1	shaker-1	7
sh-2	shaker-2	11
sha	shaven	15
shi	shiverer (= Mbp, myelin basic protein)	18
shm	shambling	11
Shmt	serine hydroxymethyl transferase	RE
sho	shorthead	UN
si	silver	10
Sig	sightless	6
Sinc	scrapie incubation period	UN
Sip	schedule-induced polydipsia	UN
Sis	simian sarcoma virus oncogene	15
Sk	scaly (SD)	UN
Skn-1	skin antigen-1 (formerly Sk-1)	UN
Skn-2	skin antigen-2 (formerly Sk-2)	UN
Sl	steel	10
sla	sex-linked anemia	X
Slf	sex-linked fidget	X
sll-1	sex-linked lethal-1 (provisional)	X
sll-2	sex-linked lethal-2 (provisional)	X
Slp	sex-limited protein within H-2S	17
slt	slaty	14
sm	syndactylism	12
smc	spondylo-metaphyseal chondrodysplasia	UN
Smg-1	submaxillary protein-1 (provisional)	7
Smg-2	submaxillary protein-2, acidic salivary protease	7
Snst	somatostatin	16
sno	snubnose	4
Soa	sucrose octaacetate aversion	UN
soc	soft coat	3
Sod-1	superoxide dismutase-1	16
Sod-2	superoxide dismutase-2	17
Sp	splotch	1
spa	spastic	3
spc	sparse coat	14
spf	sparse-fur	X
sph	spherocytosis	1
sph-(H)	spha:rozytose (provisional)	UN
Spl	plasma serotonin level	UN
Spl-1	spleen antigen-1	UN
spm	sphingomyelinosis	1
Spna-1	alpha-spectrin-1	UN
sps	spontaneous seizure (formerly dd)	2
sr	spinner	UN
Sr-1	anti-insulin antibody (proposed symbol)	2
Src-1	Rous sarcoma virus oncogene	UN
Srlv-1	susceptibility to RadLV-1	UN
srn	siren	17
srv	sensitivity to RADLV	UN
Ss	a serum beta-globulin (= C4)	17
Ssp	sex-limited saliva pattern	2
sst	short-tail shaker	UN
Sta	autosomal striping (SD)	UN
stb	stubby	X
Stk	stem cell kinetics (provisional)	UN
stm	stumpy	X
Str	striated	UN
Sts	steriod sulfatase	UN
stu	stumbler	9
su	surdescens	2
sv	Snell's waltzer	7
Svp-1	seminal vesicle protein-1	2
Svp-2	seminal vesicle protein-2	15
Svp-3	seminal vesicle protein-3	UN
sw	swaying	X
Swl	sprawling	UN
Sxa	sex chromosome association (provisional)	UN
Sxr	sex reversed	X
Sxv	susceptibility to xenotropic virus (prob. = Rmc-1)	1
sy	shaker-with-syndactylism	18
T	brachyury	17
Ta	tabby	X
Tag	temporal alpha-galactosidase	UN
Tam-1	tosyl arginine methylesterase-1	7
Tas	T-associated sex reversal	17
tb	tumbler	1
Tb	tibialess (provisional)	UN
Tbg	thyroxin binding globulin	RE
tc	truncate	6
Tcd-1	transmission distortion locus-1 (in t complex)	17
Tcd-2	transmission distortion locus-2 (in t complex)	17
Tcn-2	transcobalamin-2	11
Tcp-1	t-complex protein-1 (formerly Tp63, p63/6.9)	17
Tcp-2	t-complex protein-2	17
Tcp-3	t-complex protein-3	17
Tcp-4	t-complex protein-4	17
Tcp-5	t-complex protein-5	17
Tcp-6	t-complex protein-6	17
Tcp-7	t-complex protein-7	17
Tcp-8	t-complex protein-8	17
Tcp-9	t-complex protein-9	17
Tcr-1	transmission distortion, responder locus (in t complex)	17
Td	tattered	X
tda	testis-determining autosomal-1 (provisional)	X
Tdy	Y-linked testis-determining gene	Y
Tem	temporal control of H-2 expression (not linked to H-2, 20cM from Rec)	UN
ter	teratoma	UN
tf	tufted	17
Tfm	testicular feminization	X

Symbol	Description	Chr
tg	tottering	8
th	tilted head	1
Thb	ThB cell surface antigen	2
thf	thin fur	17
Thy-1	thymus cell antigen-1(theta)	9
Thy-2	thymus cell antigen-2	17
ti	tipsy	11
Tind	T-cell alloantigen, peripheral T cells	12
tk	tail-kinks	9
Tk-1	thymidine kinase-1	11
tl	nonerupted teeth	UN
Tla	thymus leukemia antigen	17
tn	teetering	11
Tol-1	tolerance to BGG	UN
tor	tortured	UN
tp	taupe	7
Tpi-1	triosephosphate isomerase-1	6
Tpre	pre-T-cell alloantigen	12
Tr	trembler	11
Trf	transferrin	9
trm	tremor	UN
Try-1	trypsin-1 (may = Prt-1 or Prt-3)	6
Tsa	thyrotropin-stimulating hormone, alpha subunit	11
Tsk	tight-skin	4
Tsu	T-suppressor cell alloantigen	2
Tsz-1	thymus size-1 (provisional)	12
Tthy	thymocyte alloantigen	UN
tu	toe-ulnar	
tub	tubby	7
Tw	twirler	18
twi	twitcher	UN
twt	twister	7
tx	toxic milk	UN
ty	trembly	X
U	umbrous (SD)	UN
Udpk	uridine diphosphate kinase	RE
Udpp	uridyl diphosphate glucose pyrophosphorylase	RE
Ulp	ulnaless	UN
Umph-1	uridine monophosphatase-1 (provisional)	UN
Umph-2	uridine monophosphatase-2 (provisional)	UN
Umpk	uridine monophosphate kinase	RE
un	undulated	2
Up	umbrous-patterned (provisional)	UN
Upg-1	urinary pepsinogen-1	17
Upg-2	urinary pepsinogen-2	1
Ups	uroporphyrinogen I synthase	9
us	urogenital syndrome	UN
uw	underwhite	15
v	waltzer	10
Va	varitint-waddler	3
van	viable anemia (provisional)	15
vb	vibrator	11
vc	vacillans	4
ve	velvet coat	15
vi	visceral inversion	UN
vl	vacuolated lens	1
Vlm	vacuolated lens with microphthalmia (provisional)	UN
vt	vestigial-tail	11
w	dominant spotting	5
wa-1	waved-1	6
wa-2	waved-2	11
Wap	whey acidic protein	UN
War	warfarin resistance	7
Wc	waved coat	14
wd	waddler	4
we	wellhaarig	2
wf	wavy fur (provisional)	UN
wh	writher	4
wi	whirler	UN
wl	wabbler-lethal	14
wr	wobbler	2
wst	wasted	UN
Wt	waltzer-type (SD)	UN
wuf	white underfur	
wv	weaver	16
Xce	X-chromosome controlling element	X
xid	X-linked immune deficiency	X
Xid-1	xylose dehydrogenase-1	7
xn	exencephaly	4
Xp	xeroderma pigmentosum (provisional)	X
Xpl	X-linked polydactyly	X
Xt	extra toes	13
Yaa	acelerated autoimmunity and lymphoproliferation	Y
Ym	yellow mottled	X
z	lethal (provisional)	2

Appendix 3:
Suggested Reading

Austin, C.R. and R.V. Short. 1982. *Reproduction in mammals.* Book 1. Germ cells and fertilization; Book 2. Embryonic and fetal development, 2nd edition; Book 3. Hormonal control of reproduction, 2nd edition, 1984. Cambridge University Press, England.

Daniel, J.C., ed. 1978. *Methods in mammalian reproduction.* Academic Press, New York.

Fitzgerald, M.J.T. 1978. *Human embryology.* Harper and Row, New York.

Foster, H.L., J.D. Small, and J.G. Fox, eds. 1983. *The mouse in biomedical research,* vols. I–III. Academic Press, New York.

Glasser, S.R. and D.W. Bullock, eds. 1981. *Cellular and molecular aspects of implantation.* Plenum Press, New York.

Green, E.L., ed. 1975. *Biology of the laboratory mouse,* 2nd ed. Dover Publications, New York.

Green, M.C., ed. 1981. *Genetic variants and strains of the laboratory mouse.* Gustav Fisher Verlag, Stuttgart.

Johnson, M., ed. 1977-1983. *Development in manuals,* vols. 1–4. Elsevier/North-Holland, Amsterdam.

McLaren, A. 1976. *Mammalian chimeras.* Cambridge University Press, England.

McLaren, A. and C. Wylie, eds. 1983. Current problems in germ cell differentiation. In *7th Symposium British Society of Developmental Biology.* Cambridge University Press, England.

Miller, U. and W.W. Franke. 1983. Mechanisms of gonadal differentiation in vertebrates. *Differentiation* (suppl.) *23.*

Rafferty, K.A. 1970. *Methods in experimental embryology of the mouse.* The Johns Hopkins Press, Baltimore.

Rugh, R. 1968. *The mouse: Its reproduction and development.* Burgess, Minneapolis, Minnesota.

Sherman, M.I., ed. 1977. *Concepts in mammalian embryogenesis.* MIT Press, Cambridge, Massachusetts.

Silvers, W.K. 1979. *The coat colors of mice: A model for mammalian gene action and interaction.* Springer-Verlag, New York.

Theiler, K. 1972. *The house mouse.* Springer-Verlag, Berlin.

Embryogenesis in mammals. *CIBA Found. Symp. 40,* 1976.

Embryonic and germ cell tumours in man and animals. *Cancer Surv. 2,* 1983.

New frontiers in mammalian reproduction and development. *J. Exp. Zool. 228,* 165–395, 1983.

Additional sources of information regarding genetic variants and inbred strains of mice and their genetic monitoring are found on page 16 of this manual.

References

Abramczuk, J., D. Solter, and H. Koprowski. 1977. The beneficial effect of EDTA on development of mouse one-cell embryos in chemically defined medium. *Dev. Biol. 61:* 378–383.

Adamson, E.D. 1982. The location and synthesis of transferrin in mouse embryos and teratocarcinoma cells. *Dev. Biol. 91:* 227–234.

Adamson, E.D. and S.E. Ayers. 1979. The localization and synthesis of some collagen types in developing mouse embryos. *Cell 16:* 953–965.

Adamson, E.D. and B.L.M. Hogan. 1984. Expression of EGF receptor and transferrin by F9 and PC13 teratocarcinoma cells. *Differentiation 27:* 152–157.

Altman, P.L. and D.D. Katz, eds. 1979. *Inbred and genetically defined strains of laboratory animals.* Part I: *Mouse and rat.* Federation of American Societies for Experimental Biology, Bethesda, Maryland.

Amenta, P.S., C.C. Clark, and A. Martinez-Hernandez. 1983. Deposition of fibronectin and laminin in the basement membrane of the rat parietal yolk sac: Immunohistochemical and biosynthetic studies. *J. Cell Biol. 96:* 104–111.

Anderson, W.F., L. Killos, L. Sanders-Haigh, P.J. Kretchmer, and E.G. Diakumakos. 1980. Replication and expression of thymidine kinase and human globin genes microinjected into mouse fibroblasts. *Proc. Natl. Acad. Sci. 77:* 5399–5403.

Andrews, G.K., M. Dziadek, and T. Tamaoki. 1982a. Expression and methylation of the mouse α-fetoprotein gene in embryonic, adult, and neoplastic tissues. *J. Biol. Chem. 257:* 5148–5153.

Andrews, G.K., R.G. Janzen, and T. Tamaoki. 1982b. Stability of α-fetoprotein messenger RNA in mouse yolk sac. *Dev. Biol. 89:* 111–116.

Arnheim, N., D. Treco, B. Taylor, and E.M. Eicher. 1982. Distribution of ribosomal gene length variants among mouse chromosomes. *Proc. Natl. Acad. Sci. 79:* 4677–4680.

Austin, C.R. 1961. *The mammalian egg.* Blackwell, London.

Axelrod, H.E. and E. Lader. 1983. A simplified method for obtaining embryonic stem cell lines from blastocysts. In *Cold Spring Harbor Conf. Cell Proliferation 10:* 665–670.

Balinsky, B.I. 1975. *An introduction to embryology,* 4th edition. W.B. Saunders, Philadelphia.

Barlow, D.P., N.R. Green, M. Kurkinen, and B.L.M. Hogan. 1984. Sequencing of laminin B chain cDNAs reveal C-terminal regions of coiled-coil alpha helix. *EMBO J. 3:* 2355–2362.

Beddington, R.S.P. 1981. An autoradiographic analysis of the potency of embryonic ectoderm in the 8th day postimplantation mouse embryo. *J. Embryol. Exp. Morphol. 64:* 87–104.

Bennett, K.L., R.E. Hill, D.F. Pietras, M. Woodworth-Gutai, C. Kane-Kass, J.M. Houston, J.K. Heath, and N.D. Hastie. 1984. Most highly repeated dispersed DNA families in the mouse genome. *Mol. Cell. Biol. 4:* 1561–1571.

Bensaude, O. and M. Morange. 1983. Spontaneous high expression of heat shock proteins in mouse embryonal carcinoma cells and ectoderm from day 8 mouse embryos. *EMBO J. 2:* 173–178.

Bensaude, O., C. Babinet, M. Morange, and F. Jacob. 1983. Heat shock proteins, first major products of zygotic gene activity in mouse embryo. *Nature 305:* 331–333.

Bergstrom, S. 1978. Experimentally delayed transplantation. In *Methods in mammalian reproduction* (ed. J.C. Daniel), pp. 419–435. Academic Press, New York.

Bernstine, E.G., M.L. Hooper, S. Grandchamp,

and B. Ephrussi. 1973. Alkaline phosphatase activity in mouse teratoma. *Proc. Natl. Acad. Sci.* 70: 3899–3903.

Berry, R.J., ed. 1981. *Biology of the house mouse.* The Zoological Society of London, Academic Press.

Biggers, J.D., W.K. Whitten, and D.G. Whittingham. 1971. The culture of mouse embryos in vitro. In *Methods in mammalian embryology* (ed. J.C. Daniel), pp. 86–116. W.H. Freeman, San Francisco.

Bishop, C.E., P. Boursot, B. Baron, F. Bonhomme, and D. Hatat. 1985. Most classical *mus musculus domesticus* laboratory mouse strains carry a *mus musculus musculus* Y chromosome. *Nature 315:* 70–72.

Bleil, J.D. and P.M. Wassarman. 1980a. Structure and function of the zona pellucida: Identification and characterisation of the proteins of the mouse oocyte zona pellucida. *Dev. Biol. 76:* 185–202.

———. 1980b. Synthesis of zona pellucida proteins by denuded and follicle-enclosed mouse oocytes during culture in vitro. *Proc. Natl. Acad. Sci.* 77: 1029–1033.

———. 1983. Sperm-egg interactions in the mouse: Sequence of events and induction of the acrosome reaction by a zona pellucida glycoprotein. *Dev. Biol. 95:* 317–324.

Boller, K. and R. Kemler. 1983. In vitro differentiation of embryonal carcinoma cells characterized by monoclonal antibodies against cell markers. *Cold Spring Harbor Conf. Cell Proliferation 10:* 39–49.

Bonhomme, F., U. Catalan, J. Britton-Davidian, V.M. Chapman, K. Moriwaki, E. Nevo, and L. Thaler. 1984. Biochemical diversity and evolution in the genus *Mus. Biochem. Genet.* 22: 275–303.

Boshier, D.P. 1968. The relationship between genotype and reproductive performance before parturition in mice. *J. Reprod. Fertil. 15:* 427–435.

Bouche, J.P. 1981. The effect of spermidine on endonuclease inhibition by agarose contaminants. *Anal. Biochem. 115:* 42–45.

Bradley, A., M. Evans, M.H. Kaufman, and E. Robertson. 1984. Formation of germ-line chimaeras from embryo-derived teratocarcinoma cell lines. *Nature 309:* 255–256.

Brandt, E.J., R.T. Swank, and E.K. Novak. 1981. The murine Chediak-Higashi mutation and other pigmentation mutations. In *Immunologic defects in laboratory animals* (ed. G.M.E. Gershwin and B. Merchant), vol. 1, pp. 99–117. Plenum Press, New York.

Brinster, R.L. 1965. Studies on the development of mouse embryos *in vitro.* II. The effect of energy source. *J. Exp. Zool. 158:* 59–68.

———. 1974. The effect of cells transferred into mouse blastocyst on subsequent development. *J. Exp. Med. 140:* 1049–1056.

Brinster, R.L., H.Y. Chen, M.E. Trumbauer, and M.R. Avarbock. 1980. Translation of globin mRNA by the mouse ovum. *Nature 283:* 499–501.

Brinster, R.L., H.Y. Chen, M.E. Trumbauer, and B.V. Paynton. 1981a. Secretion of proteins by the fertilized mouse ovum. *Exp. Cell Res. 134:* 291–296.

Brinster, R.L., H.Y. Chen, M. Trumbauer, A.W. Senear, R. Warren, and R.D. Palmiter. 1981b. Somatic expression of herpes thymidine kinase in mice following injection of a fusion gene into eggs. *Cell 27:* 223–231.

Brinster, R.L., H.Y. Chen, R. Warren, A. Sarthy, and R.D. Palmiter. 1982. Regulation of metallothionein-thymidine kinase fusion plasmids injected into mouse eggs. *Nature 296:* 39–42.

Brinster, R.L., H.Y. Chen, A. Messing, T. van Dyke, A.J. Levine, and R.D. Palmiter. 1984. Transgenic mice harboring SV40 T-antigen genes develop characteristic brain tumors. *Cell 37:* 359–365.

Brinster, R.L., H.Y. Chen, M.E. Trumbauer, M.K. Yagle, and R.D. Palmiter. 1985. Factors affecting the efficiency of introducing foreign DNA into mice by microinjecting eggs. *Proc. Natl. Acad. Sci. 82:* 4438–4442.

Brûlet, P., H. Condamine, and F. Jacob. 1985. Spatial distribution of transcripts of the long repeated ETn sequence during early mouse embryogenesis. *Proc. Natl. Acad. Sci. 82:* 2054–2058.

Bulfield, G. 1985. Mitochondrial DNA and house mouse speciation. *Trends Genet. 1:* 39.

Capecchi, M.R. 1980. High efficiency transformation by direct microinjection of DNA into cultured mammalian cells. *Cell 22:* 479–488.

Cascio, S.M. and P.M. Wassarman. 1982. Program of early development in the mammal: Post-transcriptional control of a class of proteins synthesised by mouse oocytes and early embryos. *Dev. Biol. 89:* 397–408.

Castle, W.E. and G.M. Allen. 1903. The heredity of albinism. *Proc. Am. Acad. Arts Sci. 38:* 603–621.

Chada, K., J. Magram, K. Raphael, G. Radice, E. Lacy, and F. Costantini. 1985. Specific expression of a foreign beta-globin gene in erythroid cells of transgenic mice. *Nature 314:* 377-380.

Champlin, A.K., D.L. Dorr, and A.H. Gates. 1973. Determining the stage of the estrous cycle in the mouse by the appearance of the vagina. *Biol. Reprod. 8:* 491–494.

Chapman, V.M., R.G. Kratzer, and B.A. Quarantillo. 1983. Electrophoretic variation for X-chromosome-linked hypoxanthine phosphoribosyl transferase (HPRT) in wild-derived mice. *Genetics 10:* 785–795.

Chapman, V., L. Forrester, J. Sanford, N. Hastie, and J. Rossant. 1984. Cell lineage specific undermethylation of mouse repetitive DNA. *Nature 307:* 284–286.

Clark, C.C., J. Crossland, G. Kaplan, and A. Martinez-Hernandez. 1982. Location and identification of the collagen found in the 14.5 d rat embryo visceral yolk sac. *J. Cell Biol. 93:* 251–260.

Clark, J.M. and E.M. Eddy. 1975. Fine structural observations on the origin and association of primordial germ cells of the mouse. *Dev. Biol. 47:* 136–155.

Clark, R. 1984. *J.B.S. The life and work of J.B.S. Haldane.* Oxford University Press, England.

Clegg, K.B. and L. Piko. 1983. Poly(A) length, cytoplasmic adenylation and synthesis of poly(A)⁺RNA in early mouse embryos. *Dev. Biol. 95:* 331–341.

Cohen, A. and M. Schlesinger. 1970. Absorption of guinea pig serum with agar. *Transplantation 10:* 130–132.

Cooper, A.R. and H. MacQueen. 1983. Subunits of laminin are differentially synthesised in mouse eggs and early embryos. *Dev. Biol. 96:* 467–471.

Cooper, A.R., M. Kurkinen, A. Taylor, and B.L.M. Hogan. 1981. Studies on the biosynthesis of laminin by murine parietal endoderm cells. *Eur. J. Biochem. 119:* 189–197.

Copeland, N.G., N.A. Jenkins, and B.K. Lee. 1983. Association of the lethal yellow (Aʸ coat color mutation with an ecotropic murine leukaemia virus genome. *Proc. Natl. Acad. Sci. 80:* 247–249.

Costantini, F. and E. Lacy. 1981. Introduction of a rabbit beta-globin gene into the mouse germ line. *Nature 294:* 92–94.

Cox, K.H., D.V. DeLeon, L.M. Angerer, and R.C. Angerer. 1984. Detection of mRNAs in sea urchin embryos by *in situ* hybridization using asymmetric RNA probes. *Dev. Biol. 101:* 485–502.

Cronmiller, C. and B. Mintz. 1978. Karyotypic normalcy and quasi-normalcy of developmentally totipotent mouse teratocarcinoma cells. *Dev. Biol. 67:* 465–477.

Cruz, Y.P. and R.A. Pedersen. 1985. Cell fate in the polar trophectoderm of mouse blastocysts as studied by microinjection of cell lineage tracers. *Dev. Biol. 112:* 73–83.

Curran, T., A.D. Miller, L. Zokas, and I.M. Verma. 1984. Viral and cellular *fos* proteins: A comparative analysis. *Cell 36:* 259–268.

Damsky, C.H., J. Richa, D. Solter, K. Korudsen, and C.A. Buck. 1983. Identification and purification of a cell surface glycoprotein mediating intracellular adhesion in embryonic and adult tissue. *Cell 34:* 455–466.

Darmon, M., J. Bottenstein, and G. Sato. 1981. Neural differentiation following culture of embryonal carcinoma cells in a serum-free defined medium. *Dev. Biol. 85:* 463–473.

De Felici, M. and A. McLaren. 1983. In vitro culture of mouse primordial germ cells. *Exp. Cell Res. 144:* 417–427.

Deol, M.S. and G.M. Truslove. 1981. Non-random distribution of unpigmented melanocytes in the retina of chinchilla-mottled mice and its significance in phenotype expression of pigment cells. In *Proceedings of the XIth International Pigment Cell Conference,* Sendai, Japan (ed. M. Seiji), pp. 153–157. University of Tokyo Press, Japan.

Derman, E. 1981. Isolation of a cDNA clone for mouse urinary proteins: Age and sex-related expression of mouse urinary protein genes is transcriptionally controlled. *Proc. Natl. Acad. Sci. 78:* 5425–5429.

Derman, E., K. Krauter, L. Walling, C. Weinberger, M. Ray, and J.E. Darnell. 1981. Transcriptional control in the production of liver–specific mRNAs. *Cell 23:* 731–739.

Dewey, M.J., D.W. Martin, Jr., G.R. Martin, and B. Mintz. 1977. Mosaic mice with teratocarcinoma-derived mutant cells deficient in hypoxanthine phosphoribosyltransferase. *Proc. Natl. Acad. Sci. 74:* 5564–5568.

Dickmann, Z. 1971. Egg transfer. In *Methods in mammalian embryology* (ed. J.C. Daniel), pp. 133–145. W.H. Freeman, San Francisco.

Diwan, S.B. and L.C. Stevens. 1976. Development of teratomas from the ectoderm of mouse egg cylinders. *J. Natl. Cancer Inst. 57:* 937–939.

Dziadek, M. 1978. Modulation of alphafoetoprotein synthesis in the early postimplantation mouse embryo. *J. Embryol. Exp. Morphol. 46:* 135–146.

———. 1981. Use of glucine as a non-enzymatic procedure for separation of mouse embryonic tissues and dissociation of cells. *Exp. Cell Res. 133:* 383–393.

Dziadek, M. and E. Adamson. 1978. Localisation and synthesis of alphafoetoprotein in post-implantation mouse embryos. *J. Embryol. Exp. Morphol. 43:* 289–313.

Dziadek, M.A. and G.K. Andrews. 1983. Tissue specificity of alpha-fetoprotein messenger RNA expression during mouse embryogenesis. *EMBO J. 2:* 549–554.

Eddy, E.M. and A.C Hahnel. 1983. Establishment of the germ cell line in mammals. In *7th Symposium of British Society for Developmental Biology* (ed. A. McLaren and C.C. Wylie), pp. 41–69. Cambridge University Press, England.

Edelman, G.M. 1985. Cell adhesion and the molecular processes of morphogenesis. *Annu. Rev. Biochem. 54:* 135–170.

Eiferman, F.A., P.R. Young, R.W. Scott, and S.M. Tilghman. 1981. Intragenic amplification and divergence in the mouse gene. *Nature 294:* 713–718.

Enders, A.C., R.L. Given, and S. Schlafke. 1978. Differentiation and migration of endoderm in the rat and mouse at implantation. *Anat. Rec. 190:* 65–78.

Enders, A.C., P.J. Chavez, and S. Schlafke. 1981. Comparison of implantation in utero and in vitro. In *Cellular and molecular aspects of implantation* (ed. S.R. Glasser and D.W. Bullock), pp. 365–382. Plenum Press, New York.

Epstein, C.J., S. Smith, B. Travis, and G. Ticker. 1978. Both X chromosomes function before visible X-chromosome inactivation in female mouse embryos. *Nature 274:* 500–503.

Epstein, C.P. 1985. Mouse monosomies and trisomies as experimental systems for studying mammalian aneuploidy. *Trends Genet. 1:* 129–134.

Evans, E.P. 1981. Karyotype of the house mouse. *Symp. Zool. Soc. Lond. 47:* 127–139.

Evans, M.J. and M.H. Kaufman. 1981. Establishment or culture of pluripotential cells from mouse embryos. *Nature 292:* 154–156.

———. 1983. Pluripotential cells grown directly from normal mouse embryos. *Cancer Surv. 2:* 185–207.

Faddy, M.J., R.G. Gosden, and R.G. Edwards. 1983. Ovarian follicle dynamics in mice: A comparative study of three inbred strains and a F1 hybrid. *J. Endocrinol. 96:* 23–24.

Ferris, S.D., R.D. Sage, and A.C. Wilson. 1982. Evidence from mt DNA sequences that common laboratory strains of inbred mice are descended from a single female. *Nature 295:* 163–165.

Festing, W.F.W. 1979. *Inbred strains in biomedical research*. Macmillan, London.

Finn, C.A. 1971. The biology of decidual cells. *Adv. Reprod. Physiol. 5:* 1–26.

Fischer, E.M., J.S. Cavanna, and S.D.M. Brown. 1985. The microdissection and microcloning of the mouse X chromosome. *Proc. Natl. Acad. Sci. 82:* 5846–5849.

Flach, G., M.H. Johnson, P.R. Braude, R.A.S. Taylor, and V.N. Bolton. 1982. The transition from maternal to embryonic control in the 2-cell mouse embryo. *EMBO J. 1:* 681–686.

Fleischman, R. and B. Mintz. 1979. Prevention of genetic anemias in mice by microinjection of normal hematopoietic stem cells into the fetal placenta. *Proc. Natl. Acad. Sci. 76:* 5736–5740.

Foster, H.L., J.D. Small, and J.G. Fox, eds. 1983. *The mouse in biomedical research*, vols. I–III. Academic Press, New York.

Franke, W.W., C. Grund, C. Kuhn, B.W. Jackson, and K. Illmensee. 1982. Formation of cytoskeletal elements during mouse embryogenesis. *Differentiation 23:* 43–59.

Fraser, L.R. and L.M. Drury. 1975. The relationship between sperm concentration and fertilization in vitro of mouse eggs. *Biol. Reprod. 13:* 513–518.

Frischauf, A.-M. 1985. The T/t complex of the mouse. *Trends Genet. 1:* 100–103.

Fujii, J.T. and G.R. Martin. 1983. Developmental potential of teratocarcinoma stem cells in utero following aggregation with cleavage stage mouse embryos. *J. Embryol. Exp. Morphol. 74:* 79–96.

Gardner, R.L. 1968. Mouse chimeras obtained by the injection of cells into the blastocyst. *Nature 220:* 596–597.

———. 1971. Manipulations on the blastocyst. *Adv. Biosci. 6:* 279–296.

———. 1982. Investigation of cell lineage and differentiation in the extraembryonic endoderm of the mouse embryo. *J. Embryol. Exp. Morphol. 68:* 175–198.

———. 1983. Origin and differentiation of extra-embryonic tissues in the mouse. *Int. Rev. Exp. Pathol. 24:* 63–133.

———. 1984. An *in situ* cell marker for clonal analysis of development of the extraembryonic endoderm in the mouse. *J. Embryol. Exp. Morphol. 80:* 251–288.

Gardner, R.L. and J. Rossant. 1979. Investigation of the fate of 4.5 d post coitum mouse ICM cells by blastocyst injection. *J. Embryol. Exp. Morphol. 52:* 141–152.

Gates, A.H. 1971. Maximizing yield and developmental uniformity of eggs. In *Methods in mammalian embryology* (ed. J.C. Daniel), pp. 64–76. W.H. Freeman, San Francisco.

Gautsch, J.W. and M.C. Wilson. 1983. Delayed de novo methylation in teratocarcinoma suggests additional tissue-specific mechanisms for controlling gene expression. *Nature 301:* 32–37.

Gerhard, D.S., E.S. Kawasaki, F.C. Bancroft, and P. Szabo. 1981. Localization of a unique gene by direct hybridization *in situ*. *Proc. Natl. Acad. Sci. 78:* 3755-3759.

Gielbelhaus, D.H., J.J. Heikkila, and G.A. Schultz. 1983. Changes in the quantity of histone and actin mRNA during the development of preimplantation mouse embryos. *Dev. Biol. 98:* 148–154.

Giles, R.E. and F.H. Ruddle. 1973. Production of Sendai virus for cell fusion. *In Vitro 9:* 103–107.

Giloh, H. and J.W. Sedat. 1982. Fluorescence microscopy: Reduced photobleaching of rhodamine and fluorescein protein conjugates by *n-propyl* gallate. *Science 217:* 1252–1255.

Glass, R.H., J. Aggeler, A. Spindle, R.A. Pederson, and Z. Werb. 1983. Degradation of extracellular matrix by mouse trophoblast outgrowths: A model for implantation. *J. Cell Biol. 96:* 1108–1116.

Gonda, M.A. and Y.-C. Hsu. 1980. Correlative scanning electron, transmission electron, and light microscopic studies of the in vitro development of mouse embryos on a plastic substrate at the implantation stage. *J. Embryol. Exp. Morphol. 56:* 23–39.

Gordon, J.W. and F.H. Ruddle. 1981. Integration and stable germ line transmission of genes injected into mouse pronuclei. *Science 214:* 1244–1246.

Gordon, J.W., G.A. Scangos, D.J. Plotkin, J.A. Barbosa, and F.H. Ruddle. 1980. Genetic transformation of mouse embryos by microinjection of purified DNA. *Proc. Natl. Acad. Sci. 77:* 7380–7384.

Gorer, P.A., S. Lyman, and G.D. Snell. 1948. Studies on the genetic and antigenic basis of tumour transplantation. Linkage between a histocompatibility gene and "fused" in mice. *Proc. R. Soc. Lond. B 135:* 499–505.

Gorin, M.B. and S.M. Tilghman. 1980. Structure of the gene in the mouse. *Proc. Natl. Acad. Sci. 77:* 1351–1355.

Graham, C.F. 1971. Virus assisted fusion of embryonic cells. *Acta Endocrinol.* (suppl.) *153:* 154–167.

Graham, C.F. and Z.A. Deussen. 1978. Features of cell lineage in preimplantation mouse embryos. *J. Embryol. Exp. Morphol. 48:* 53–72.

Green, E.L., ed. 1975. *Biology of the laboratory mouse,* 2nd edition. Dover, New York.

Green, M.C., ed. 1981. *Genetic variants and strains of the laboratory mouse.* Gustav Fischer Verlag, Stuttgart.

Greve, J.M. and P.M. Wassarman. 1985. Mouse egg extracellular coat is a matrix of interconnected filaments possessing a structural repeat. *J. Mol. Biol. 181:* 253–264.

Gropp, A. and H. Winking. 1981. Robertsonian translocations: Cytology, meiosis, segregation patterns and biological consequences of heterozygosity. *Symp. Zool. Soc. Lond. 47:* 141–181.

Grosschedl, R., D. Weaver, D. Baltimore, and F. Costantini. 1984. Introduction of a μ immunoglobulin gene into the mouse germ line: Specific expression in lymphoid cells and synthesis of functional antibody. *Cell 38:* 647–658.

Grüneberg, H. 1947. *Animal genetics and medicine.* Hamish Hamilton, London.

———. 1952. *The genetics of the mouse,* 2nd edition. Martinus Nijhoff, The Hague.

———. 1961. Genetic studies on the skeleton of the mouse. XXIX. Pudgy. *Genet. Res. 2:* 384–393.

———. 1963. *The pathology of development: A study of inherited skeletal disorders in mammals.* Blackwell, Oxford, England.

Hadorn, E. 1961. *Developmental genetics and lethal factors.* Methuen, London.

Hafen, E., M. Levine, R.L. Garber, and W.J. Gehring. 1983. An improved *in situ* hybridization method for the detection of cellular RNAs in *Drosophila* tissue sections and its application for localizing transcripts of the homeotic *Antennapedia* gene complex. *EMBO J. 2:* 617–623.

Haldane, J.B.S., A.D. Sprunt, and N.M. Haldane. 1915. Reduplication in mice. *J. Genet. 5:* 133–135.

Hammond, J. 1949. Recovery and culture of tubal mouse ova. *Nature 163:* 28–29.

Hanahan, D. 1985. Heritable formation of pancreatic beta-cell tumors in transgenic mice expression recombinant insulin/simian virus 40 oncogenes. *Nature 315:* 115–122.

Harbers, K., D. Jahner, and R. Jaenisch. 1981. Microinjection of cloned retroviral genomes into mouse zygotes: Integration and expression in the animal. *Nature 293:* 540–542.

Hatta, K. and M. Takeichi. 1986. Expression of N-cadherin adhesion molecules associated with early morphogenic events in chick development. *Nature 320:* 447–449.

Heape, W. 1890. Preliminary note on the transplantation and growth of mammalian ova within a uterine foster mother. *Proc. R. Soc.*

Lond. B 48: 457.

Heath, J.K. 1983. Regulation of murine embryonal carcinoma cell proliferation and differentiation. *Cancer Surv. 2:* 141–164.

Heiniger, H.-J. and J.J. Dorey. 1980. *Handbook on genetically standard Jax mice.* The Jackson Laboratory, Bar Harbor, Maine.

Herrman, B., M. Bucan, P. Mains, A.-M. Frischauf, L.M. Silver, and H. Lehrach. 1986. Genetic analysis of the proximal portion of the mouse *t* complex: Evidence for a second inversion within *t* haplotypes. *Cell 44:* 469–476.

Hogan, B.L.M. 1980. High molecular weight extracellular proteins synthesized by endoderm cells derived from mouse teratocarcinoma cells and normal extraembryonic membranes. *Dev. Biol. 76:* 275–285.

Hogan, B.L.M. and P.R. Cooper. 1982. Synthesis of Reichert's membrane components by parietal endoderm cells of the mouse embryo. In *New trends in basement membrane research* (ed. K. Kuehn et al.), pp. 245–255. Raven Press, New York.

Hogan, B.L.M. and R. Tilly. 1977. In vitro culture and differentiation of normal mouse blastocysts. *Nature 265:* 626–629.

———. 1978a. In vitro development of inner cell masses isolated immunosurgically from mouse blastocysts. II. Inner cell masses from 3.5–4.0-day p.c. blastocysts. *J. Embryol. Exp. Morphol. 45:* 107–121.

———. 1978b. In vitro development of inner cell masses isolated immunosurgically from mouse blastocysts. *J. Embryol. Exp. Morphol. 45:* 93–121.

———. 1981. Cell interactions and endoderm differentiation in cultured mouse embryos. *J. Embryol. Exp. Morphol. 62:* 379–394.

Hogan, B.L.M., A.R. Cooper, and M. Kurkinen. 1980. Incorporation into Reichert's membrane of laminin-like extracellular proteins synthesized by parietal endoderm cells of the mouse embryos. *Dev. Biol. 80:* 289–300.

Hogan, B.L.M., A. Taylor, and E.D. Adamson. 1981. Cell interaction modulates embryonal carcinoma cell differentiation into parietal visceral endoderm. *Nature 291:* 235–237.

Hogan, B.L.M., D.P. Barlow, and R. Tilly. 1983. F9 teratocarcinoma cells as a model for the differentiation of parietal and visceral endoderm in the mouse embryo. *Cancer Surv. 2:* 115–140.

Hogan, B.L.M., D.P. Barlow, and M. Kurkinen. 1984. Reichert's membrane as a model system for biosynthesis of basement membrane components. *CIBA Found. Symp. 108:* 60–69.

Hogan, B.L.M., P. Holland, and P. Schofield. 1985.

How is the mouse segmented? *Trends Genet. 1:* 67–74.

Hoppe, P.C. and K. Illmensee. 1977. Microsurgically produced homozygous-diploid uniparental mice. *Proc. Natl. Acad. Sci. 74:* 5657–5661.

———. 1981. Full-term development after transplantation of parthenogenetic embryonic nucleic into fertilized mouse eggs. *Proc. Natl. Acad. Sci. 79:* 1912–1916.

Hoppe, P.C. and S. Pitts. 1973. Fertilization in vitro and development of mouse ova. *Biol. Reprod. 8:* 420–426.

Howlett, S.K. and V.N. Bolton. 1985. Sequence and regulation of morphological and molecular events during the first cell cycle of mouse embryogenesis. *J. Embryol. Exp. Morphol. 87:* 175–206.

Hsu, Y.-C. 1978. In vitro development of whole mouse embryos beyond the implantation stage. In *Methods in mammalian reproduction* (ed. J.C. Daniel), pp. 229–245. Academic Press, New York.

———. 1980. Embryo growth and differentiation factors in embryonic sera of mammals. *Dev. Biol. 76:* 465–474.

Hsueh, A.J.W., E.Y. Adashi, P.B.C. Jones, and T.H. Welsh. 1984. Hormonal regulation of the differentiation of cultured ovarian granulosa cells. *Endocrine Rev. 5:* 76–127.

Huang, T.T.F. and P.G. Calarco. 1981. Evidence for the cell surface expression of intracisternal A particle-associated antigens during early mouse development. *Dev. Biol. 82:* 388–392.

Hyafil, F., D. Morello, C. Babinet, and F. Jacob. 1980. A cell surface glycoprotein involved in the compaction of embryonal carcinoma cells and cleavage stage embryos. *Cell 21:* 927–934.

Hyafil, F., C. Babinet, and F. Jacob. 1981. Cell-cell interactions in early embryogenesis: A molecular approach to the role of calcium. *Cell 26:* 447–454.

Ilgren, E.B. 1981. Homotypic cellular interactions in mouse trophoblast development. *J. Embryol. Exp. Morphol. 62:* 183–202.

Illmensee, K. and P.C. Hoppe. 1981. Nuclear transplantation in *Mus musculus:* Developmental potential of nuclei from preimplantation embryos. *Cell 23:* 9–18.

Ingham, P.W., K.R. Howard, and D. Ish-Horowicz. 1985. Transcription pattern of the *Drosophila* segmentation gene *Hairy. Nature 318:* 439–445.

Iyengar, M.R., C.W.L. Iyengar, H.Y. Chen, R.L. Brinster, E. Bornslaeger, and R.M. Schultz.

1983. Expression of creatine kinase isoenzyme during oogenesis and embryogenesis in the mouse. *Dev. Biol. 96:* 263–268.

Izant, J.G. and H. Weintraub. 1985. Constitutive and conditional suppression of exogenous and endogenous genes by anti-sense RNA. *Science 229:* 345–352.

Jackson, B.W., C. Grund, E. Schmid, K. Burki, W.W. Franke, and K. Illmensee. 1980. Formation of cytoskeletal elements during mouse embryogenesis. I. Intermediate filaments of the cytokeratin type and desmosomes in preimplantation embryos. *Differentiation 17:* 161–179.

Jackson, B.W., C. Grund, S. Winter, W.W. Franke, and K. Illmensee. 1981. Formation of cytoskeletal elements during mouse embryogenesis. II. Epithelial differentiation and intermediate-sized filaments in early postimplantation embryos. *Differentiation 20:* 203–216.

Jacob, F. 1983. Concluding remarks. *Cold Spring Harbor Conf. Cell Proliferation 10:* 683–687.

Jaenisch, R. 1976. Germ line integration and Mendelian transmission of the exogenous Moloney leukemia virus. *Proc. Natl. Acad. Sci. 73:* 1260–1264.

———. 1980. Retroviruses and embryogenesis: Mircoinjection of Moloney leukemia virus into midgestation mouse embryos. *Cell 19:* 181–188.

———. 1985. Mammalian neural crest cells participate in normal embryonic development on microinjection into post-implantation mouse embryos. *Nature 318:* 181–183.

Jaenisch, R. and B. Mintz. 1974. Simian virus 40 DNA sequences in DNA of healthy adult mice derived from preimplantation blastocysts injected with viral DNA. *Proc. Natl. Acad. Sci. 71:* 1250–1254.

Jaenisch, R., D. Jahner, P. Nobis, I. Simon, J. Lohler, K. Harbers, and D. Grotkopp. 1981. Chromosomal position and activation of retroviral genomes inserted into the germ line of mice. *Cell 24:* 519–529.

Jaenisch, R., K. Harbers, A. Schnieke, J. Lohler, I. Chumakov, D. Jahner, D. Grotkopp, and E. Hoffman. 1983. Germline integration of Moloney murine leukemia virus at the Mov13 locus leads to recessive lethal mutation and early embryonic death. *Cell 32:* 209–216.

Jahner, D., H. Stuhlman, C.L. Stewart, K. Harbers, J. Lohler, I. Simon, and R. Jaenisch. 1982. De novo methylation and expression of retroviral genomes during mouse embryogenesis. *Nature 298:* 623–628.

Jahner, D., K. Haase, R. Mulligan, and R. Jaen-

isch. 1985. Insertion of the bacterial *gpt* gene into the germ line of mice by retroviral infection. *Proc. Natl. Acad. Sci. 82:* 6927–6931.

Janzen, R.G., G.L. Andrews, and T. Tamaoki. 1982. Synthesis of secretory proteins in developing mouse yolk sac. *Dev. Biol. 90:* 18–23.

Jenkins, N.A., N.G. Copeland, B.A. Taylor, and B.K. Lee. 1981. Dilute (d) coat colour mutation of DBA/2J mice is associated with the site of integration of an ecotropic MuLV genome. *Nature 293:* 370–374.

John, M., E. Carswell, E.A. Boyse, and G. Alexander. 1972. Production of θ antibody by mice that fail to reject θ-incompatible skin grafts. *Nat. New Biol. 238:* 57–58.

Johnson, G.D., R.S. Davidson, K.C. McNamee, G. Russell, D. Goodwin, and E.J. Holborow. 1982. Fading of immunofluorescence during microscopy: A study of the phenomenon and its remedy. *J. Immunol. Methods 55:* 231–242.

Johnson, M.H. 1981. The molecular and cellular basis of preimplantation mouse development. *Biol. Rev. 56:* 463–498.

Johnson, M.H. and C.A. Ziomek. 1981. The foundation of two distinct cell lineages within the mouse morula. *Cell 24:* 71–80.

Johnson, M.H., H.P.M. Pratt, and A.H. Handyside. 1981. The generation and recognition of positional information in the preimplantation mouse embryo. In *Cellular and molecular aspects of implantation* (ed. S.R. Glassner and D.W. Bullock), pp. 55–73. Plenum Press, New York.

Kafatos, F.C., W.C. Jones, and A. Efstratiadis. 1979. Determination of nucleic acid sequence homologies and relative concentrations by a dot hybridization procedure. *Nucleic Acids Res. 7:* 1541–1552.

Karfunkel, P. 1974. The mechanism of neural tube formation. *Int. Rev. Cytol. 38:* 245–271.

Kaufman, M.H. 1978a. The experimental production of mammalian parthogenetic embryos. In *Methods in mammalian reproduction* (ed. J.C. Daniel), pp. 21–47. Academic Press, New York.

———. 1978b. The chromosome complement of single pronuclear haploid mouse embryos following activation by ethanol treatment. *J. Embryol. Exp. Morphol. 71:* 139–154.

———. 1981. Parthogenesis: A system facilitating understanding of factors that influence early mammalian development. *Prog. Anat. 1:* 1–34.

———. 1982. The chromosome complement of single-pronuclear haploid mouse embryos following activation by ethanol treatment. *J.*

Embryol. Exp. Morphol. 71: 139–154.

———. 1983a. Early mammalian development: Parthenogenetic studies. *Dev. Cell Biol. 14.*

———. 1983b. Ethanol-induced chromosomal abnormalities at conception. *Nature 302:* 258–260.

Kaufman, M.H., S.C. Barton, and M.A.H. Surani. 1977. Normal post-implantation development of mouse parthenogenetic embryos to the forelimb bud stage. *Nature 265:* 53–55.

Kaufman, M.H., E.J. Robertson, A.H. Handyside, and M.J. Evans. 1983. Establishment of pluripotential cell lines from haploid mouse embryos. *J. Embryol. Exp. Morphol. 73:* 249–261.

Keeler, C. 1978. How it began. In *Origins of inbred mice* (ed. H.C. Morse), pp. 179–192. Academic Press, New York.

Kelly, S.J. 1977. Studies on the development potential of 4- and 8-cell stage mouse blastomeres. *J. Exp. Zool. 200:* 365–376.

Kemler, R., C. Babinet, H. Eisen, and F. Jacobs. 1977. Surface antigen in early differentiation. *Proc. Natl. Acad. Sci. 74:* 4449–4452.

Kemler, R., R. Brûlet, M.-T. Schnebelen, J. Gaillard, and F. Jacob. 1981. Reactivity of monoclonal antibodies against intermediate filament proteins during embryonic development. *J. Embryol. Exp. Morphol. 64:* 45–60.

King, W., M.D. Patel, L.I. Lobel, S.P. Goff, and M.C. Nguyen-Huu. 1985. Insertion mutagenesis of embryonal carcinoma cells by retroviruses. *Science 228:* 554–558.

Klein, J. 1975. *Biology of the mouse histocompatibility-2 complex.* Springer-Verlag, Berlin.

Kratzer, P.G., V.M. Chapman, H. Lambert, R.E. Evans, and R.M. Liskay. 1983. Differences in the DNA of the inactive X chromosomes of fetal and extraembryonic tissues of mice. *Cell 33:* 37–42.

Krco, C.J. and E.H. Goldberg. 1976. H-Y (male) antigen detection on eight cell mouse embryos. *Science 193:* 1134–1135.

Krumlauf, R., R.E. Hammer, S.M. Tilghman, and R.L. Brinster. 1985. Developmental regulation of alpha-fetoprotein in transgenic mice. *Mol. Cell. Biol. 5:* 163–168.

Krumlauf, R., V. Chapman, R. Hammer, R. Brinster, and S.M. Tilghman. 1986. Differential expression of alpha-fetoprotein genes on the inactive X chromosome in extraembryonic and somatic tissues of a transgenic mouse line. *Nature 319:* 224–226.

Kurkinen, M., D.P. Barlow, L. Foster, and B.L.M. Hogan. 1982. In vitro synthesis of type IV procollagen. *J. Biol. Chem. 257:* 15151–15155.

Kurkinen, M., D.P. Barlow, D. Helfman, J.G. William, and B.L.M. Hogan. 1983a. cDNAs for basement membrane components. Type IV collagen. *Nucleic Acid Res. 11:* 6199–6209.

Kurkinen, M., D.P. Barlow, J.R. Jenkins, and B.L.M. Hogan. 1983b. In vitro synthesis of laminin and entactin polypeptides. *J. Biol. Chem. 258:* 6543–6548.

Labarca, C. and K. Paigen. 1980. A simple, rapid and sensitive DNA assay procedure. *Anal. Biochem. 102:* 344–352.

Lacy, E., S. Roberts, E.P. Evans, M.D. Burtenshaw, and F. Costantini. 1983. A foreign beta-globin gene in transgenic mice: Integration at abnormal chromosomal positions and expression in inappropriate tissues. *Cell 34:* 343–358.

Lane, E.B., B.L.M. Hogan, M. Kurkinen, and J.I. Garrels. 1983. Coexpression of vimentin and cytokeratins in parietal endoderm cells of the early mouse embryo. *Nature 303:* 701–704.

Le Douarin, N. 1982. *The neural crest.* Cambridge University Press, England.

Lehtonen, E., V.-P. Lehto, R. Paasivuo, and I. Virtanen. 1983a. Parietal and visceral endoderm differ in their expression of intermediate filaments. *EMBO J. 2:* 1023–1028.

Lehtonen, E., V.-P. Lehto, T. Vartio, R.A. Badley, and I. Virtanen. 1983b. Expression of cytokeratin polypeptides in mouse oocytes and preimplantation embryos. *Dev. Biol. 100:* 158–165.

Leibo, P. and P. Mazur. 1978. Methods for the preservation of mammalian embryos by freezing. In *Methods in mammalian reproduction* (ed. J.C. Daniel), pp. 179–201. Academic Press, New York.

Leivo, I., A. Vaheri, R. Timpl, and J. Wartiovaara. 1980. Appearance and distribution of collagens and laminin in the early mouse embryo. *Dev. Biol. 76:* 100–114.

Levak-Svajger, B., A. Svajger, and N. Skreb. 1969. Separation of serum layers in presomite rat embryos. *Experientia 25:* 1311–1312.

Lewis, N.E. and J. Rossant. 1982. Mechanism of size regulation in mouse embryo aggregates. *J. Embryol. Exp. Morphol. 72:* 169–181.

Lewis, W.H. and P.W. Gregory. 1929. Cinematographs of living developing rabbit eggs. *Science 69:* 226–229.

Lin, T.P., J. Florence, and O. Jo. 1973. Cell fusion induced by a virus within the zona pellucida of mouse eggs. *Nature 242:* 47–49.

Lobel, L.I., M. Patel, W. King, M.C. Nguyen-Huu, and S.P. Goff. 1985. Construction and recovery of viable retroviral genomes carrying a bacterial suppressor transfer RNA gene. *Science 228:* 329–331.

Lohler, J., R. Timpl, and R. Jaenisch. 1984. Embryonic lethal mutation in mouse collagen I gene causes rupture of blood vessels and is associated with erythropoietic and mesenchymal cell death. *Cell 38:* 597–607.

Loutit, J.F. and B.M. Cattenach. 1983. Haematopoietic role for Patch (P*h)* revealed by new W mutant (W^(ct)) in mice. *Genet. Res. 42:* 23–39.

Lyon, M.F. 1961. Gene action in the X-chromosome of the mouse. *Nature 190:* 372–373.

Lyon, M.F. and S. Rastan. 1984. Parental source of chromosome implantation and its relevance for X-chromosome inactivation. *Differentiation 26:* 63–67.

Lyon, M.F., E.P. Evans, S.E. Jarvis, and I. Sayers. 1979. *t* haplotypes of the mouse may involve a change in intercalary DNA. *Nature 279:* 38–42.

MacQueen, H.A. 1979. Lethality of radioisotopes in early mouse embryos. *J. Embryol. Exp. Morphol. 52:* 203–208.

Magram, J., K. Chada, and F. Constantini. 1985. Developmental regulation of a cloned adult beta-globin gene in transgenic mice. *Nature 315:* 338–340.

Maniatis, T., E.F. Fritsch, and J. Sambrook. 1982. *Molecular cloning: A laboratory manual.* Cold Spring Harbor Laboratory, Cold Spring Harbor, New York.

Mark, W., K. Signorelli, and E. Lacy. 1985. A recessive lethal mutation in a transgenic mouse line. *Cold Spring Harbor Symp. Quant. Biol. 50:* 447–452.

Markert, C.L. 1982. Partheneogenesis, homozygosity, and cloning in mammals. *J. Hered. 73:* 390–397.

Maro, B., M.H. Johnson, S.J. Pickering, and G. Flach. 1984. Changes in actin distribution during fertilization of the mouse egg. *J. Embryol. Exp. Morphol. 81:* 211–237.

Marotti, K.R., D. Belin, and S. Strickland. 1982. The production of distinct forms of plasminogen activator by mouse embryonic cells. *Dev. Biol. 90:* 154–159.

Martin, G.R. 1980. Teratocarcinoma and mammalian embryogenesis. *Science 209:* 768–776.

———. 1981. Isolation of a pluripotent cell line from early mouse embryos cultured in medium conditioned by teratocarcinoma stem cells. *Proc. Natl. Acad. Sci. 78:* 7634–7636.

Martin, G.R. and M.J. Evans. 1975. Differentiation of clonal lines of teratocarcinoma cells: Formation of embryoid bodies in vitro. *Proc. Natl. Acad. Sci. 72:* 1441–1445.

Mayer, T.C. 1973. Site of gene action in Steel mice: Analysis of the pigment defect by mesodermal-ectodermal recombinations. *J. Exp. Zool. 184:* 345–352.

Mayr, E. 1982. *The growth of biological thought, diversity, evolution and inheritance.* The Belknap Press of Harvard University Press, Cambridge, Massachusetts.

McBurney, M.W. and B.J. Rogers. 1982. Isolation of male embryonal carcinoma cells lines and their chromosome replication patterns. *Dev. Biol. 89:* 503–508.

McBurney, M.W., E.M.V. Jones-Villeneuva, M.K.S. Edwards, and P.J. Anderson. 1982. Control of muscle and neuronal differentiation in a cultured embryonal carcinoma cell line. *Nature 299:* 165–167.

McCoshen, J.A. and D.J. McCallion. 1975. A study of the primordial germ cells during their migratory phase in Steel mutant mice. *Experientia 31:* 589–590.

McGinnis, W., R.L. Garber, J. Wirz, A. Kuroiwa, and W.J. Gehring. 1984. A homologous protein-coding sequence in *Drosophila* homeotic genes and its conservation in other metazoans. *Cell 27:* 403–408.

McGrath, J. and D. Solter. 1983a. Nuclear transplantation in mouse embryos. *J. Exp. Zool. 228:* 355–362.

———. 1983b. Nuclear transplantation in the mouse embryo by microsurgery and cell fusion. *Science 220:* 1300–1302.

———. 1984a. Maternal T hp lethality in the mouse is a nuclear, not cytoplasmic defect. *Nature 308:* 550–551.

———. 1984b. Completion of mouse embryogenesis requires both the maternal and paternal genomes. *Cell 37:* 179–183.

———. 1984c. Inability of mouse blastomere nuclei transferred to enucleated zygotes to support development *in vitro. Science 226:* 1317–1319.

McKusick, V.A. and F.H. Ruddle. 1977. The status of the gene map of the human chromosomes. *Science 196:* 390–405.

McLaren, A. 1976. *Mammalian chimeras.* Cambridge University Press, England.

———. 1983. Does the chromosomal sex of a mouse germ cell affect its development. In *7th Symposium of British Society for Developmental Biology* (ed. A. McLaren and C.C. Wylie), pp. 225–240. Cambridge University Press, England.

McLaren, A. and J.D. Biggers. 1958. Successful development and broth of mice cultivated in vitro as early embryos. *Nature 182:* 877–878.

McLaren, A. and D. Michie. 1956. Studies on the

transfer of fertilized mouse eggs to uterine foster-mothers. I. Factors affecting the implantation and survival of native and transferred eggs. *J. Exp. Biol. 33:* 394–416.

McMahon, A., M. Fosten, and M. Monk. 1983. X-chromosome inactivation mosaicism in the three germ layers and the germ line of the mouse embryo. *J. Embryol. Exp. Morphol. 74:* 207–220.

Meehan, R.R., D.P. Barlow, R.E. Hill, B.L.M. Hogan, and N.D. Hastie. 1984. Pattern of serum protein gene expression in mouse visceral yolk sac and fetal liver. *EMBO J. 3:* 1881–1885.

Meier, S. 1979. Development of the chick embryo mesoblast. Formation of the embryonic axis and establishment of the metameric pattern. *Dev. Biol. 73:* 25–45.

———. 1981. Development of the chick embryo mesoblast: Morphogenesis of the prechordal plate and cranial segments. *Dev. Biol. 83:* 49–61.

Meier, S. and P.P.L. Tam. 1982. Metameric pattern development in the embryonic axis of the mouse. I. Differentiation of the cranial segments. *Differentiation 21:* 95–108.

Melton, D.A., P.A. Krieg, M.R. Rebagliati, T. Maniatis, K. Zinn, and M.R. Green. 1984. Efficient *in vitro* synthesis of biologically active RNA and RNA hybridization probes from plasmids containing a bacteriophage SP6 promoter. *Nucleic Acids Res. 12:* 7035–7056.

Miller, D.A. and O.J. Miller. 1981. Cytogenetics. In *The mouse in biomedical research* (ed. H.L. Foster et al.), vol. 1, pp. 241–261. Academic Press, New York.

Mintz, B. 1964. Formation of genetically mosaic mouse embryos, and early development of "lethal (t12/t12)-normal" mosaics. *J. Exp. Zool. 157:* 273–292.

———. 1967. Mammalian embryo culture. In *Methods in developmental biology* (ed. E.H. Wilt and N.K. Wessels), pp. 379–400. Cromwell, New York.

———. 1971. Allophenic mice of multiembryo origin. In *Methods in mammalian embryology* (ed. J.C. Daniel), pp. 186–214. W.H. Freeman, San Francisco.

———. 1978. Gene expression in neoplasia and differentiation. *Harvey Lect. 71:* 193–245.

Mintz, B. and K. Illmensee. 1975. Normal genetically mosaic mice produced from malignant teratocarcinoma cells. *Proc. Natl. Acad. Sci. 72:* 3585–3589.

Mintz, B. and E.S. Russell. 1957. Gene-induced embryological modifications of primordial germ cells in the mouse. *J. Exp. Zool. 134:* 207–237.

Monk, M. and A. McLaren. 1981. X-chromosome activity in foetal germ cells of the mouse. *J. Embryol. Exp. Morphol. 63:* 75–84.

Morris, R.J. and P.C. Barber. 1983. Fixation of Thy-1 in nervous tissue for immunochemistry. *J. Histol. Cytochem. 31:* 263–274.

Morriss-Kay, G.M. 1981. Growth and development of pattern in the cranial neural epithelium of rat embryos during neurulation. *J. Embryol. Exp. Morphol.* (suppl.) *65:* 225–241.

Morse, H.C. 1978. *Origins of inbred mice.* Academic Press, New York.

———. 1981. The laboratory mouse—A historical perspective. In *The mouse in biomedical research. History, genetics and wild mice* (ed. H.L. Foster et al.), vol. 1, pp. 1–16. Academic Press, New York.

Muggleton-Harris, A., D.G. Whittingham, and L. Wilson. 1982. Cytoplasmic control of preimplantation development in vitro in the mouse. *Nature 299:* 460–462.

Muglia, L. and J. Locker. 1984. Extrapancreatic insulin gene expression in the fetal rat. *Proc. Natl. Acad. Sci. 81:* 3635–3639.

Mullen, R.J. 1977. Site of *Ped* gene action and Purkinje cell mosaicism in cerebella of chimeric mice. *Nature 270:* 245–247.

Muller, R., I.M. Verma, and E.D. Adamson. 1983. Expression of c-*onc* genes: c-*fos* transcripts accumulate to high levels during development of mouse placenta, yolk sac and amnion. *EMBO J. 2:* 679–684.

Munke, M., K. Harbers, R. Jaenisch, D.R. Cox, and U. Franke. 1985. Assignment of the alpha-1(I) collagen gene to mouse chromosome II. *Cytogenet. Cell Genet.* (in press).

Nadijcka, M. and N. Hillman. 1974. Ultrastructural studies of the mouse blastocyst substages. *J. Embryol. Exp. Morphol. 32:* 675–695.

Needham, J. 1959. *A history of embryology.* Cambridge University Press, England.

Neff, J.M. and J.F. Enders. 1968. Poliovirus replication and cytogenicity in monolayer hamster cell cultures fused with beta propiolactone-inactivated Sendai virus. *Proc. Soc. Exp. Biol. Med. 127:* 260–267.

New, D.A.T. 1978. Whole-embryo culture and the study of mammalian embryos during organogenesis. *Biol. Rev. 53:* 81–122.

Nicolet, G. 1970. Analyse autoradiographique de la localisation des differentes ebauches presomptives dans la ligne primitive l'embryon de poulet. *J. Embryol. Exp. Morphol. 23:* 79–108.

Nielsen, J.T. and V.M. Chapman. 1977. Electrophoretic variation for X-chromosome-linked phosphoglycerate kinase (PGK-1) in the mouse. *Genetics 87:* 319–325.

Niwa, O., Y. Yokota, H. Ishida, and T. Sugahara. 1983. Independent mechanisms involved in supression of the Moloney leukemia virus genome during differentiation of murine teratocarcinoma cells. *Cell 32:* 1105-1113.

Noden, D.M. 1983. Embryonic origins of avian cephalic and cervical muscles and associated connective tissues. *Am. J. Anat. 168:* 257–276.

Nomura, T., N. Ohsawa, N. Tamaoki, and K. Fujiura, eds. 1977. *Proceedings of the 2nd International Workshop on Nude Mice.* University of Tokyo Press, Japan.

Nomura, T., K. Esaki, and T. Tomita. 1985. *ICLAS manual for genetic monitoring of inbred mice.* University of Tokyo Press, Japan.

Oppenheimer, J.M. 1967. *Essays on the history of embryology and biology.* The MIT Press, Cambridge, Massachusetts.

Oshima, R.G., W.H. Howe, J.M. Tabor, and K. Trevor. 1983. Cytoskeletal proteins as markers of embryonal carcinoma differentiation. *Cold Spring Harbor Conf. Cell Proliferation 10:* 51–61.

Overbeek, P.A., A.B. Chepelinsky, J.S. Khillan, J. Piatigorsky, and H. Westphal. 1985. Lens-specific expression and developmental regulation of the bacterial chloramphenicol acetyltransferase gene driven by the murine alpha A-chrystallin promoter in transgenic mice. *Proc. Natl. Acad. Sci. 82:* 7815–7819.

Palmiter, R.D. and R.L. Brinster. 1985. Transgenic mice. *Cell 41:* 343–345.

Palmiter, R.D., R.L. Brinster, R.E. Hammer, M.E. Trumbauer, M.G. Rosenfeld, N.C. Birnberg, and R.M. Evans. 1982. Dramatic growth of mice that develop from eggs microinjected with a metallothionein-growth hormone fusion gene. *Nature 300:* 611–615.

Papaioannou, V.E., M.W. McBurney, and R.L. Gardner. 1975. Fate of teratocarcinoma cells injected into early mouse embryos. *Nature 258:* 70.

Papaioannou, V.E., R.L. Gardner, M.W. McBurney, C. Babinet, and M.J. Evans. 1978. Participation of cultured teratocarcinoma cells in mouse embryogenesis. *J. Embryol. Exp. Morphol. 44:* 93–104.

Paynton, B.V., K.M. Ebert, and R.L. Brinster. 1983. Synthesis and secretion of ovalbumin by mouse growing oocytes following microinjection of chick ovalbumin mRNA. *Exp. Cell Res. 144:* 214–218.

Peters, J., ed. 1985. *Mouse news letter 72,* p. 27. Oxford University Press, England.

Peyrieras, N., F. Hyafil, D. Louvard, H.L. Ploegh, and F. Jacob. 1983. Uvomorulin, a non-integral membrane protein of early mouse embryo. *Proc. Natl. Acad. Sci. 80:* 6274–6277.

Phillips, S.J., E.H. Birkenmeier, R. Callahan, and E.M. Eicher. 1982. Male and female mouse DNAs can be discriminated using retroviral probes. *Nature 297:* 241–243.

Pienkowski, M., D. Solter, and H. Koprowski. 1974. Early mouse embryos: Growth and differentiation in vitro. *Exp. Cell Res. 85:* 424–428.

Pincus, G. 1936. *The eggs of mammals.* Macmillan, New York.

Ponder, B.A.T., M.M. Wilkinson, and M. Wood. 1983. H2 antigens as a marker of cellular genotype in chimeric mice. *J. Embryol. Exp. Morphol. 76:* 83–93.

Potten, C.S. 1985. *Radiation and skin.* Taylor and Francis, London.

Pratt, H.P.M., J. Chakraborty, and M.A.H. Surani. 1981. Molecular and morphological differentiation of the mouse blastocyst after manipulation of compaction with cytochalasin D. *Cell 26:* 279–292.

Pratt, H.P.M., V.N. Bolton, and K.A. Gudgeon. 1983. The legacy from the oocyte and its role in controlling early development of the mouse embryo. *CIBA Found. Symp. 98:* 197–227.

Quinn, P., C. Barros, and D.G. Whittingham. 1982. Preservation of hamster oocytes to assay the fertilizing capacity of human spermatozoa. *J. Reprod. Fertil. 66:* 161–168.

Rafferty, R.A. 1970. *Methods in experimental embryology of the mouse.* The Johns Hopkins Press, Baltimore.

Rastan, S. 1982. Timing of X-chromosome inactivation in post-implantation mouse embryos. *J. Embryol. Exp. Morphol. 71:* 11–24.

Reeve, W.J.D. and C.A. Ziomek. 1981. Distribution of microvilli on dissociated blastomeres from mouse embryos: Evidence for surface polarization at compaction. *J. Embryol. Exp. Morphol. 62:* 339–350.

Rinchik, E.M., L.B. Russell, N.G. Copeland, and N.A. Jenkins. 1985. The dilute-short ear (d-se) complex of the mouse: Lessons from a fancy mutation. *Trends Genet. 1:* 170–176.

Rizzino, A. and C. Crowley. 1980. The growth and differentiation of the ECC line F9 in defined media. *Proc. Natl. Acad. Sci. 77:* 457–461.

Rizzino, A. and M.I. Sherman. 1979. Development and differentiation of mouse blastocysts in serum-free medium. *Exp. Cell Res. 121:* 221–223.

Roach, A., N. Takahashi, D. Pravtcheva, F. Ruddle, and L. Hood. 1985. Chromosomal mapping of mouse myelin basic protein gene and structure and transcription of the partially deleted gene in shiverer mutant mice. *Cell 42:* 149–155.

Robert, B., P. Barton, A. Minty, P. Daubas, A. Weydert, F. Bonhomme, J. Catalan, D. Chazottes, J.-L. Cuenet, and M. Buckingham. 1985. Investigation of genetic linkage between myosin and actin genes using an interspecific mouse backcross. *Nature 314:* 181–183.

Robertson, E.J., M.J. Evans, and M.H. Kaufman. 1983. X chromosome instability in pluripotential stem cell lines derived from parthogenic embryos. *J. Embryol. Exp. Morphol. 74:* 297–309.

Robins, D.M., S. Ripley, A.S. Henderson, and R. Axel. 1981. Transforming DNA integrates into the host chromosome. *Cell 23:* 29–39.

Rodriguez, J. and F. Deinhardt. 1960. Preparation of a semipermanent mounting medium for fluorescent antibody studies. *Virology 12:* 316–217.

Roger, A.W. 1979. *Techniques of autoradiography,* 3rd edition. Elsevier/North-Holland, Amsterdam.

Rossant, J. and B.A. Croy. 1985. Genetic identification of tissue of origin of cellular populations within the mouse placenta. *J. Embryol. Exp. Morphol. 86:* 177–189.

Rossant, J. and V.E. Papaioannou. 1977. The biology of embryogenesis. In *Concepts in mammalian embryogenesis* (ed. M.I. Sherman), pp. 1–36. The MIT Press, Cambridge, Massachusetts.

Rossant, J., K.M. Vijh, C.E. Grossi, and M.D. Cooper. 1986. Clonal origin of hematopoietic colonies in the postnatal mouse liver. *Nature 319:* 507–510.

Rubenstein, J.L.R., J.-F. Nicolas, and F. Jacob. 1984. Construction of a retrovirus capable of transducing and expressing genes in multipotential embryonic cells. *Proc. Natl. Acad. Sci. 81:* 7137–7140.

Rubin, G.M. and A.C. Spradling. 1982. Genetic transformation of *Drosophila* with transposable element vectors. *Science 218:* 348–353.

Russell, E.S. 1954. One man's influence: A tribute to William Ernest Castle. *J. Hered. 45:* 210–213.

———. 1978. Origins and history of mouse inbred strains: Contribution of Clarence Cook Little. In *Origins of inbred mice* (ed. H.C. Morse), pp. 33–43. Academic Press, New York.

Russell, L.B. 1971. Deprection of functional units in a small chromosomal segment of the mouse and its use in interpreting the nature of radiation induced mutations. *Mutat. Res. 11:* 107–123.

Rutishauser, U. and C. Goridis. 1986. NCAM: The molecule and its genetics. *Trends Genet. 2:* 72–76.

Sanford, J., L. Forrester, V. Chapman, A. Chandley, and N. Hastie. 1984. Methylation of repetitive DNA sequences in germ cells of *M. musculus. Nucleic Acids Res. 12:* 2823–2836.

Sawicki, J.A., T. Magnuson, and C.J. Epstein. 1981. Evidence for expression of the paternal genome in the two-cell mouse embryo. *Nature 295:* 450–451.

Schatten, G., C. Simmerly, and H. Schatten. 1985. Microtubule configuration during fertilization, mitosis, and early development in the mouse and the requirement for egg microtubule mediated motility during mammalian fertilization. *Proc. Natl. Acad. Sci. 82:* 4152–4156.

Schmidt, G.H., M.M. Wilkinson, and B.A. Ponder. 1985. Cell migration pathway in the intestinal epithelium: An *in situ* marker system using mouse aggregation chimeras. *Cell 40:* 425–429.

Schnieke, A., K. Harbers, and R. Jaenisch. 1983. Embryonic lethal mutation in mice induced by retrovirus insertion into the alpha1(I) collagen gene. *Nature 304:* 315–320.

Schultz, R.M., G.E. Letourneau, and P.M. Wasserman. 1979. Program of early development in the mammal. Changes in the patterns and absolute rates of tubulin and total protein synthesis during oocyte growth in the mouse. *Dev. Biol. 73:* 120–133.

Scott, R.W., T.F. Vogt, M.E. Croke, and S.M. Tilgham. 1984. Tissue-specific activation of a cloned alpha-foetoprotein gene during differentiation of a transfected carcinoma cell line. *Nature 310:* 562–567.

Semoff, S., B.L.M. Hogan, and C.R. Hopkins. 1982. Localisation of fibronectin, laminin-entactin and entactin in Reichert's membrane by immunoelectron microscopy. *EMBO J. 1:* 1171–1175.

Shani, M. 1985. Tissue-specific expression of rat myosin light-chain 2 gene in transgenic mice. *Nature 314:* 283–286.

Shimkin, M.B. 1975. A.E.C. Lathrop (1868–1918). Mouse woman of Granby. *Cancer Res. 35:* 1597–1598.

Shine, I. and S. Wrobel. 1976. *Thomas Hunt Morgan. Pioneer of genetics.* pp. 61. University of Kentucky Press, Lexington.

Shirayoshi, Y., T.S. Okada, and M. Takeichi. 1983. The Ca++ dependent cell-cell adhesion system regulates inner cell mass formation and cell surface polarization in early mouse development. *Cell 35:* 631–638.

Silver, L.M. 1981. Genetic organization of the mouse t complex. *Cell 27:* 239–240.

———. 1986. Mouse t haplotypes. *Annu. Rev. Genet. 19:* 179–208.

Silver, L.M., G.R. Martin, and S. Strickland, eds. 1983. Teratocarcinoma stem cells. *Cold Spring Harbor Conf. Cell Proliferation,* vol. 10.

Silvers, W.K. 1979. *The coat colors of mice: A model for mammalian gene action and interaction.* Springer-Verlag, New York.

Siracusa, L.D., V.M. Chapman, K.L. Bennett, N.D. Hastie, D.F. Pietras, and J. Rossant. 1983. Use of repetitive DNA sequences to distinguish *Mus musculus* and *Mus caroli* cells by in situ hybridization. *J. Embryol. Exp. Morphol. 73:* 163–178.

Slack, J.M.W. 1983. *From egg to embryo.* Cambridge University Press, England.

Smith, K.K. and S. Strickland. 1981. Structural components and characteristics of Reichert's membrane, an extraembryonic basement membrane. *J. Biol. Chem. 256:* 4654–4661.

Smith, L.J. 1985. Embryonic axis orientation in the mouse and its correlation with blastocyst relationships to the uterus. II. Relationships from 4 1/2 to 9 1/2 days. *J. Embryol. Exp. Morphol. 89:* 15–35.

Snell, G.D. 1978. Congenic resistant strains of mice. In *Origins of inbred mice* (ed. H.C. Morse), pp. 119–155. Academic Press, New York.

Snow, M.H.L. 1977. Gastrulation in the mouse: Growth and regionalization of the epiblast. *J. Embryol. Exp. Morphol. 42:* 293–303.

———. 1978a. Proliferative centres in embryonic development. In *Development in mammals* (ed. M.H. Johnson), vol. 3, pp. 337–362. Elsevier/North-Holland, Amsterdam.

———. 1978b. Techniques for separating early embryonic tissues. In *Methods in mammalian reproduction* (ed. J.C. Daniel), pp. 167–178. Academic Press, New York.

———. 1981. Autonomous development of parts isolated from primitive-streak-stage mouse embryos. Is development clonal? *J. Embryol. Exp. Morphol.* (suppl.) *65:* 269–287.

Snow, M.H.L. and M. Monk. 1983. Emergence and migration of mouse primordial germ cells. In *Current problems in germ cell differentiation* (ed. A. McLaren and C.C. Wylie), pp. 115–136. Cambridge University Press, England.

Soares, M.J., J.A. Julian, and S.R. Glasser. 1985. Trophoblast giant cell release of placental lactogens: Temporal and regional characteristics. *Dev. Biol. 107:* 520–526.

Solter, D. and B.B. Knowles. 1975. Immunosurgery of mouse blastocysts. *Proc. Natl. Acad. Sci. 72:* 5099–5102.

Southern, E. 1975. Detection of specific sequences among DNA fragments separated by gel electrophoresis. *J. Mol. Biol. 98:* 503–517.

Stevens, L.C. 1970. The development of transplantable teratocarcinomas from intratesticular grafts of pre- and post-implantation mouse embryos. *Dev. Biol. 21:* 364–382.

Stevens, L.C. and C.C. Little. 1954. Spontaneous testicular teratomas in an inbred strain of mice. *Proc. Natl. Acad. Sci. 40:* 1080–1087.

Stevens, L.C. and G.B. Pierce. 1975. Teratomas: Definitions and terminology. In *Teratomas and differentiation* (ed. M.I. Sherman and D. Solter), pp. 13–16. Academic Press, New York.

Stevens, L.C., D.S. Varnum, and E.M. Eicher. 1977. Viable chimaeras produced from normal and parthenogenetic mouse embryos. *Nature 269:* 515–517.

Stewart, C.L. 1982. Formation of viable chimeras by aggregation between teratocarcinomas and preimplantation mouse embryos. *J. Embryol. Exp. Morphol. 67:* 167–179.

Stewart, T.A. and B. Mintz. 1982. Recurrent germ-line transmission of the teratocarcinoma genome from the METT-1 culture line to progeny in vivo. *J. Exp. Zool. 224:* 465–469.

Stewart, C.L., H. Stuhlman, D. Jahner, and R. Jaenisch. 1982. De novo methylation, expression and infectivity of retroviral genomes introduced into embryonal carcinoma cells. *Proc. Natl. Acad. Sci. 79:* 4098–4102.

Storb, U., R.L. O'Brien, M.D. McMullen, K.A. Gollahon, and R.L. Brinster. 1984. High expression of cloned immunoglobulin kappa gene in transgenic mice is restricted to B lymphocytes. *Nature 310:* 238–240.

Strickland, S. and V. Mahdavi. 1978. The induction of differentiation in teratocarcinoma stem cells by retinoic acid. *Cell 15:* 393–403.

Strickland, S., K.K. Smith, and K.R. Marotti. 1980. Hormonal induction and differentiation in teratocarcinoma stem cells: Generation of parietal endoderm by retinoic acid and dibutyryl cAMP. *Cell 21:* 347–355.

Strong, L.C. 1978. Inbred mice in science. In *Origins of inbred mice* (ed. H.C. Morse), pp. 45–66. Academic Press, New York.

Stuhlmann, H., R. Cone, R.C. Mulligan, and R. Jaenisch. 1984. Introduction of a selectable gene into different animal tissue by a retrovirus recombinant vector. *Proc. Natl. Acad. Sci. 81:* 7151–7155.

Surani, M.A.H. and S.C. Barton. 1983. Development of gynogenetic eggs in the mouse: Implications for parthenogenetic embryos. *Science 222:* 1034–1036.

Surani, M.A.H., S.C. Barton, and M.H. Kaufman. 1977. Development to term of chimaeras between diploid parthenogenetic and fertilized embryos. *Nature 270:* 601–603.

Surani, M.A.H., S.C. Barton, and M.L. Norris. 1984. Development of reconstituted mouse eggs suggests imprinting of the genome during gametogenesis. *Nature 308:* 548–550.

Swift, G.H., R.E. Hammer, R.J. MacDonald, and R.L. Brinster. 1984. Tissue-specific expression of the rat pancreatic elastase I gene in transgenic mice. *Cell 38:* 639–646.

Takagi, N. and M. Sasaki. 1975. Preferential inactivation of the paternally derived X chromosome in the extraembryonic membranes of the mouse. *Nature 256:* 640–641.

Tam, P.P.L. 1981. The control of somitogenesis in mouse embryos. *J. Embryol. Exp. Morphol.* (suppl.) *65:* 103–128.

Tam, P.P.L. and S. Meier. 1982. The establishment of a somitomeric pattern in the mesoderm of the gastrulating mouse embryo. *Am. J. Anat. 164:* 209–225.

Tam, P.P.L. and M.H.L. Snow. 1980. The in vitro culture of mouse embryos. *J. Embryol. Exp. Morphol. 59:* 131–143.

———. 1981. Proliferation and migration of primordial germ cells during compensatory growth in mouse embryos. *J. Embryol. Exp. Morphol. 64:* 133–147.

Tam, P.P.L., S. Meier, and A.G. Jacobson. 1982. Differentiation of the metameric pattern in the embryonic axis of the mouse. II. Somitomeric organisation of the presomitic mesoderm. *Differentiation 21:* 109–122.

Theiler, K. 1972. *The house mouse.* Springer-Verlag, New York.

———. 1983. *Embryology in the mouse in biomedical research,* vol. 3. Academic Press, New York.

Thomas, P.S. and M.N. Farquhar. 1978. Specific measurement of DNA in nuclei and nucleic acids using diaminobenzoic acid. *Anal. Biochem. 89:* 35–44.

Townes, T.M., J.B. Lingrel, R.L. Brinster, and R.D. Palmiter. 1985. Erythroid specific expression of human beta-globin genes in transgenic mice. *EMBO J. 4:* 1715–1723.

Tuckett, F., L. Lim, and G. Morriss-Kay. 1985. The ontogenesis of cranial neuromeres in the rat embryo. II. A scanning electron microscrope and kinetic study. *J. Embryol. Exp. Morphol. 87:* 215–228.

Van Beneden, M.E. 1875. La maturation de l'oeuf, la fecondation, et les premieres phases du developpement embryonnaire des mammiferes d'apres des recherches faites chez le lapin. *Bull. Acad. R. Belg. Cl. Sci. 40:* 686–736.

Van Blerkom, J. 1981. Structural relationship and posttranslational modification of stage-specific proteins synthesised during early preimplantation development in the mouse. *Proc. Natl. Acad. Sci. 78:* 7629–7633.

van der Putten, H., F.M. Botteri, A.D. Miller, M.G. Rosenfelc, H. Fan, R.M. Evans, and I.M. Verma. 1985. Efficient insertion of genes into the mouse germ line via retroviral vectors. *Proc. Natl. Acad. Sci. 82:* 6148–6152.

Vogelstein, B. and D. Gillespie. 1979. Preparative and analytical purification of DNA from agarose. *Proc. Natl. Acad. Sci. 76:* 615–619.

Wagner, E.F., T.A. Stewart, and B. Mintz. 1981. The human beta-globin gene and a functional thymidine kinase gene in developing mice. *Proc. Natl. Acad. Sci. 78:* 5016–5020.

Wagner, E.F., L. Covarrubias, T.A. Stewart, and B. Mintz. 1983. Prenatal lethalities in mice homozygous for human growth hormone gene sequences integrated in the germ line. *Cell 35:* 647–655.

Wagner, E.F., M. Vanek, and B. Vennstrom. 1985. Transfer of genes into embryonal carcinoma cells by retrovirus infection: Efficient expression from an internal promoter. *EMBO J. 4:* 663–666.

Wagner, T.E., P.C. Hoppe, J.D. Jollick, D.R. Scholl, R.L. Hodinka, and J.B. Gault. 1981. Microinjection of a rabbit beta-globin gene into zygotes and its subsequent expression in adult mice and their offspring. *Proc. Natl. Acad. Sci. 78:* 6376–6380.

Wahl, G.M., L. Vitto, R.A. Padgett, and G.R. Start. 1982. Single-copy and amplified CAD genes in Syrian hamster chromosomes localized by a highly sensitive method for *in situ* hybridization. *Mol. Cell Biol. 2:* 308–319.

Wartiovaara, J., I. Leivo, and A. Vaheri. 1979. Expression of the cell surface-associated glycoprotein, fibronectin, in the early mouse embryo. *Dev. Biol. 69:* 247–257.

Wassarman, P.M., J.M. Greve, R.M. Perona, R.J. Roller, and G.S. Salzmann. 1984. How mouse eggs put on and take off their extracellular coat. In *Molecular biology of development* (ed. E. Davidson and R. Firtel), p. 213. A.R. Liss, New York.

Watanabe, T., M.J. Dewey, and B. Mintz. 1978.

Teratocarcinoma cells as vehicles for introducing specific mutant mitochondrial genes into mice. *Proc. Natl. Acad. Sci. 75:* 5113–5117.

Whitten, W.K. 1956. Culture of tubal mouse ova. *Nature 177:* 96.

———. 1971. Embryo medium. Nutrient requirements for the culture of preimplantation embryos in vitro. *Adv. Biosci. 6:* 129–141.

Whitten, W.K. and J.D. Biggers. 1968. Complete development *in vitro* of the preimplantation stages of the mouse in a simple chemically defined medium. *J. Reprod. Fertil. 17:* 399–401.

Whitten, W.K. and A.K. Champlin. 1978. Pheromones, estrus, ovulation and mating. In *Methods in mammalian reproduction* (ed. J.C. Daniel), pp. 403–417. Academic Press, New York.

Whittingham, D.G. 1971. Culture of mouse ova. *J. Reprod. Fertil.* (suppl.) *14:* 7–21.

———. 1980a. Principles of embryo preservation. In *Low temperature preservation in medicine and biology* (ed. A. Smith and J. Farrant), pp. 65–84. Pitman, Kent, England.

———. 1980b. Parthenogenesis in mammals. *Oxford Rev. Reprod. Biol. 2:* 205–231.

Whittingham, D.G. and M.J. Wood. 1983. Reproductive physiology in the mouse. *Biomed. Res. 111:* 137–164.

Wiley, L.M., A.I. Spindle, and R.A. Pedersen. 1978. Morphology of isolated mouse inner cell masses developing in vitro. *Dev. Biol. 63:* 1–10.

Woychik, R.P., J.A. Stewart, L.G. Davis, P. D'Eustachio, and P. Leder. 1985. An inherited limb deformity created by insertional mutagenesis in a transgenic mouse. *Nature 318:* 36–40.

Wu, T.-C., Y.-J. Wan, and I. Damjanov. 1981. Positioning of ICM determines the development of mouse blastocysts in vitro. *J. Embryol. Exp. Morphol. 65:* 105–117.

Wu, T.-C., Y.-J. Wan, A.E. Chung, and I. Damjamov. 1983. Immunohistochemical localization of entactin and laminin in mouse embryos and fetuses. *Dev. Biol. 100:* 496–505.

Wudl, L. and V. Chapman. 1976. The expression of β-glucuronidase during preimplantation development of mouse embryos. *Dev. Biol. 48:* 104–109.

Yoshida, C. and M. Takeichi. 1982. Teratocarcinoma cell adhesion: Identification of a cell-surface protein involved in calcium-dependent cell aggregation. *Cell 28:* 217–224.

Yoshida-Noro, C., N. Suzuki, and M. Takeichi. 1984. Molecular nature of the Ca^{++} dependent cell-cell adhesion system in mouse and embryonic cells studied with a monoclonal antibody. *Dev. Biol. 101:* 19–27.

Ziomek, C.A. and M.H. Johnson. 1980. Cell surface interaction induces polarization of mouse 8-cell blastomeres at compaction. *Cell 21:* 935–942.

Subject Index

A (*agouti*) locus. *See also agouti* phenotype.
  as a complex locus determining coat color, 72
  number of alleles of, 72
A^y, as a homozygous lethal mutation, 72. *See also yellow* mutation; *agouti*
Absorption, by visceral endoderm, 63–64
Acid Tyrode, for removing zona pellucida, 106
Acrosome, structure of, in sperm, 30
Acrosomal reaction, required for fertilization, 30
Actin. *See also* Cytoskeleton.
  in egg, before and after fertilization, 33
  inhibition of polymerization, by cytochalasin B, 33
  messenger RNA levels, during early development, 39
Adhesion, cell–cell. *See also* CAMs (cell adhesion molecules); Cell adhesion.
  calcium-dependent
    of compacting embryos, 42
    of F9 cells, 41
  changes in
    during compaction, 41, 42
    during gastrulation, 53
    during maturation of somites, 56, 58
  uvomorulin and, 41
Age, of embryo
  at different stages of development, 20–23
  convention used for reporting, 23
agouti (A) phenotype. See also *A* (*agouti*) locus; *Non-agouti* (a) phenotype.
  determined by mesodermal component of hair follicle, 72
  examples of mice with, 74–77
  variation in, over different parts of the body, 72
albino (c) locus. *See also albino* mutation (c).
  mutations in, affecting coat color, 73
albino mutation (c).
  effect on
    pigmentation of both eyes and coat, 73, 74

tyrosinase activity, 73, 74
  visual pathways, 73
  use of, to demonstrate Mendelian inheritance, 2, 74
Albumin, in relation to AFP, 65
Alcohol, parthenogenetic activation of oocytes by, 33, 36, 109–110
Allantois
  germ cells at base of, 19
  origin, structure and fate of, 68
  in 8.5-day p.c. embryo, 50, 123
  in relation to placenta, 70
Alkaline phosphatase
  activity in germ cells, 19, 25, 26
  histochemical staining using, 26
Alpha-fetoprotein (AFP)
  as a marker for visceral endoderm, 65, 66
  evolutionary relationship to albumin, 65
  first synthesis of, 65
  messenger RNA
    amount of, in visceral endoderm, 65
    detection of, by in situ hybridization, 65
  synthesis regulated by cell interactions, 65
  synthesis and secretion of
    by fetal or regenerating liver, 65
    by placenta, 70
    by visceral endoderm, 62, 63, 64–65
Amnion
  basement membrane in, 68, 69
  development of, 68
  dissection of, from 13.5-day p.c. embryo, 124–125
  expression of c-*fos* in, 62
  structure and position of
    in 8.5-day p.c. embryo, 50
    in 10.5-day p.c. embryo, 69
    in 13.5-day p.c. embryo, 62
Ampulla
  recovery of fertilized eggs from, 100–101
  site of fertilization, 30, 35
Analysis, of foreign DNA in transgenic mice,

317

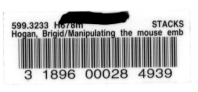